Wire EDM Handbook

by
Carl Sommer and Steve Sommer M.E.

Carl and Steve Sommer operate the largest
wire EDM company west of the Mississippi—Reliable EDM Corporation

Advance Publishing, Inc • HOUSTON

Advance Publishing , Inc.
6950 Fulton Street
Houston, TX 77022
Phone (713) 695-0600

Printed in the United States

Library of Congress Catalog Card Number 92-90878

ISBN # 1-57537-301-7

Acknowledgments

We want to thank the following individuals for reviewing the manuscript: Alan D. Simon, Applications Engineer for Japax; Terry Bryce, wire EDM machine tool sales, and Sy Roerick, Applications Engineer for T. Bryce and Associates; Marc Townsend, Vice President of United Tool and Stamping Company, West Paterson, N.J.; Tom Schweda, president of Schweda EDM Sales; and Gregory V. Langenhorst, wire EDM applications manager for Mitsubishi.

About the Authors

As the owner and president of Reliable EDM Corporation, the largest wire EDM job shop west of the Mississippi, Carl Sommer is uniquely qualified to write a book on the wire EDM process.

His introduction to the machine tool trade goes back to 1949, at a New York city machine shop. It was not long before Sommer began working as an apprentice tool and die maker where he learned to make dies with hand files and a filing machine.

During the next twenty years Sommer gained broad, valuable experience in virtually all areas of the field—precision tools and dies, fixtures, and short run production for such companies as IBM, Gyrodyne, Thikol, Fairchild Stratus, Reminton, and Sikorsky Helicopter. He operated all the machines, worked in the inspection department, and made precision dies where parts were ground to within .0001″. He eventually became a foreman in a tool and die and stamping company.

For the better part of the 1970s Sommer worked as a high school teacher in the New York City Public School System. During this time he also conducted extensive research into the problems facing America's educational institutions. This research culminated in the writing of the book, *Schools in Crisis: Training for Success or Failure?* This book has had a major impact on school reform in many states.

Sommer moved to Houston, Texas in 1978 and re-entered the machine tool industry—first as a tool and die maker, then as a tool designer for one of Houston's largest tool and die and stamping shops. He advanced to the position of operations manager, and for more than five years managed the entire company.

In December of 1986, Sommer began Reliable EDM with his two sons Steve and Phil. Within just nine years Reliable EDM became the largest wire EDM job shop west of the Mississippi River.

Other works by Sommer include a twelve-part series of mathematics books promoting adult education, an adult literacy series with an emphasis on phonics, and numerous children's picture books encouraging virtues for successful living.

Steve Sommer, vice president of Reliable EDM, received his mechanical engineering degree from the University of Houston. He has experience as a machinist and a tool and die maker, and he possesses a thorough knowledge of computer programming and the wire EDM process. He and his brother Phil manage Reliable EDM.

The exceptional experience of this father-and-sons team is one of the major reasons for Reliable's remarkable growth. Following are pictures of Reliable EDM and some of its equipment.

Reliable EDM Corporation
6940 Fulton St. • Houston, TX 77022
1-800-WIRE EDM (800-947-3336)
Tel. (713) 692-5454
FAX (713) 692-2466
http://www.reliableedm.com

Foreword

When I first saw a copy of Carl and Steve Sommer's book, The Wire EDM Handbook, I was certain it would be a "Best Seller." This professionally written handbook covers the subject of Wire EDM in such a clear, concise manner that it makes a complex process easy to understand. My first impression proved to be correct, and a second printing was soon necessary to meet the demand.

Written for people at every level of EDM, The Wire EDM Handbook puts a wealth of practical up-to-date information at your fingertips, and it belongs in every shop that employs Wire EDM. If you job EDM work out, are just starting with EDM, or are just curious about what EDM can accomplish, this is the book to read.

Never before in the history of Wire EDM has so much useful information been available in a single source, and *EDM Today* congratulates Carl and Steve Sommer on a job well done!

Jack Sebzda Sr.
Editor-in-Chief,
EDM Today

Table of Contents

Acknowledgments iii

About the Authors v

Foreword vii

Unit One: The Technological Revolution: Wire EDM

1. Fundamentals of Wire EDM

Revolutionizing Machining 19
New Demands By Design Engineers 22
Two Basic Types Of EDM 22
- A. Plunge EDM 22
- B. Wire EDM 23

How Wire EDM Works 23-24
The Step By Step EDM Process 25-26
- A. Power Supply Generates Volts and Amps 25
- B. Controlled Spark Erodes Material During On Time 25
- C. Off Time Allows Fluid to Remove Eroded Particles 26
- D. Filter Removes Chips While Cycle is Repeated 26

Super Precision Band Saw 27
Three Types Of Wire EDM 28-32
- A. Two Axis 28
- B. Simultaneous Four Axis 28
- C. Independent Four Axis 29-32

Understanding Independent Four Axis 32

2. Profiting With Wire EDM

Users of Wire EDM 35
Benefits Of Wire EDM 35-42
- A. Production Runs 35
- B. Various Shapes and Sizes 36
- C. Accuracy and Finishes 36
- D. Eliminates Extra Machining Processes 36
- E. Prototypes Economically Produced 37
- F. Burr Free and Perfectly Straight Machining 37
- G. Damaged Parts Can Be Repaired with Inserts 38
- H. Less Need for Skilled Craftsmen 38
- I. Material Hardness Not a Factor 38-39
- J. Computer Calculations 39
- K. Digitizing 40
- L. Miniaturization 40
- M. Fabrication of Graphite Electrodes 41

N. Air Machining 41
O. Reliable Repeatability 42
Parts for Wire EDM 42
A. Precision Gages and Templates 42
B. Keyways vs. Broaching 43
C. Shaft Slots 43
D. Collets 43
E. Parting Tubes and Shafts 44
F. Shaft Pockets 44
G. Punches and Dies From One Piece of Tool Steel 44
H. Short Run Stampings 45
I. Molds 45
J. Special and Production Tool Cutters 46
K. Difficult-to-Machine Shapes 46
L. Other Cost Reducing Parts 47-48
Advantages of Wire EDM vs. Plunge EDM 49
A. Producing Costly Electrodes 49
B. Designing Cost Effective Parts 50
Advantages of Wire EDM vs. Laser 50
A. Better Surface Finish and Accuracy 50
B. Cutting Shim Stock Absolutely Burr Free 51-52
Materials That Can be EDMed 52
Machining Costs 52
Entering the New World of Wire EDM Machining 53

3. Proper Procedures

Starting Methods for Edges and Holes 57-58
Edge Preparation 58-60
A. Square Edges 58
B. Scale 59
C. Pick-up 60
Starter Holes 60-62
A. Automatic Pick-up 60
B. Unsquare Holes 61
C. Relieved Holes 61
D. Smooth Holes 61
E. Placement and Location of Starter Holes 62
Layout 63

4. Understanding the Wire EDM Process

Accuracy and Tolerances 67
Finishes 67-68
Wire Path 68-69
A. Wire Kerf 68
B. Corners and Radii 69
Skim Cutting 69-72
A. Barreling Effect and Wire Trail Off 70
B. Metal Movement 71
C. Finishes and Accuracy 71
D. Carbide 72

Flushing 72-73
Cutting Speed 73
Impurities 73-74
Recast 74
Heat-Treated Steels 74
Cutting Large Sections 74-75
Cutting Sections From a Block 76
A. Leaving a Frame 76
B. Strength of Frame 76
C. Material for Clamping 77
Understanding the Wire EDM Process 77

5. Reducing Costs

Create One Slug 81-82
Cutting in the Air 83
Machining After Wire EDM 84-85
Multiple Plate and Sheet Metal Parts 85-86
Production Lots 86
Pre-machining Non-complicated Shapes 87

6. Production Wire EDM

Defining Production EDM 91-93
The Greatest Problem 93
Outsourcing or Owning a Wire EDM 93
Advantages of Job Shop Wire EDM Companies 93-96
A. Competitive Outsourcing 93
B. Quality 94
C. Service 94
D. Efficiency 94
E. Fixtures 94
F. Inspection Equipment 94
G. Machine and Operating Costs 95
H. Competitive Pricing 96
I. Altered Machining Methods 96

7. Various Wire EDM Applications

Examples of EDM Uses 99
A. Tall Parts 99-101
B. Overhanging Parts 101-102
C. Other Applications 102-104
Wave of the Future 104

Unit Two: The Tool and Die Revolution

8. Advantages of Using Wire EDM for Die Making

Old-Fashioned Tool and Die Making 109-111
The Revolution 112

Advantages of Wire EDM Dies 112-115
A. One-Piece Die Sections 112
B. Spare Parts 113
C. Dowel Holes EDMed 113
D. Better Tool Steels 113
E. Accuracy 113
F. Die Repairs 113-114
G. Fine Textured Finish 114
H. Eliminates Distortion 114
I. Inserts for High Wear Areas 114
J. Smaller Dies 114
K. Longer Lasting 114
L. Punches and Dies From One Piece of Tool Steel 115
M. Cutting Stripper and Die Section Together 115

9. Wire EDMing Punch and Die Sections

Punches 119
A. Large Punches 119
B. Small Punches 120-124
1. Footed (Figure 9:3) 120
2. Shoulder (Figure 9:4) 120
3. Keyed In--Toe Clamps (Figure 9:5,6) 121
4. Keyed In--Keyway (Figure 9:7) 121
5. Press Fit (Figure 9:8) 122
6. Peened Edge (Figure 9:9) 122
7. Dowel Pin Reamed (Figure 9:10) 122
8. Dowel Pin EDMed (Figure 9:11) 123
9. Set Screws (Figure 9:12) 123
10. Socket Head Cap Screw (Figure 9:13) 123
11. Ball Bearings (Figures 9:14,15) 124
C. Skim Cutting 125
Die Sections 125-127
A. Heavy Blanking Dies 125
B. Avoid Sharp Corners 125
C. Heat-Treating 125
D. Large Die Sections 126
E. Tapers 126-127
1. Taper and Land 126
2. Straight Cut and Taper 126-127

10. One-Piece Stamping Dies

Blanking Die 131-137
A. Desired Stamping 131
B. Preparing the Tool Steel Blank 131-132
C. Placement of Starter Hole 132
1. Punch Shape is the Desired Stamping 132-133
2. Blank Shape is the Desired Shape 133-134
D. Harden the Tool Steel Blank 134
E. Stripper Plate 134
F. Punch Holder 134
G. Mount Die Blank on Die Set 135

H. Mount Punch Holder on Die Set 135-136
I. Mount Stripper on Bottom of Die Section 136
J. Wire Cut Punch, Die and Stripper 137
Total Burr-Free One-Piece Blanking Die 138
Compound Blanking Dies 139-143
A. Desired Stamping (Figure 10:15) 139
B. Prepare Tool Steel Blank 139
C. Mount Punch Holder 140
D. Mount Die Block on Bottom of Die Set 140
E. Mount Punch Holder on Top of Die Set 141
F. Stripper Plate 141
G. Wire EDM Compound Die 141-142
H. Completed Die 142
1. Stripper with Springs 142
2. Stripper with Knockout 143
The Tool And Die Revolution 144

Appendix (Questions for Textbook Use)

Chapter 1. Fundamentals of Wire EDM 147-148
Chapter 2. Profiting With Wire EDM 148-149
Chapter 3. Proper Procedures 149
Chapter 4. Understanding the Wire EDM Process 150
Chapter 5. Reducing Costs 150-151
Chapter 6. Production Wire EDM 151
Chapter 7. Various Wire EDM Applications 151-152
Chapter 8. Advantages of Using Wire EDM for Die Making 152
Chapter 9. Wire EDMing Punch and Die Sections 152-153
Chapter 10. One-Piece Stamping Dies 153

Unit One

The Technological Revolution:

Wire EDM

The Revolution: Wire EDM

Courtesy Charmilles Technology

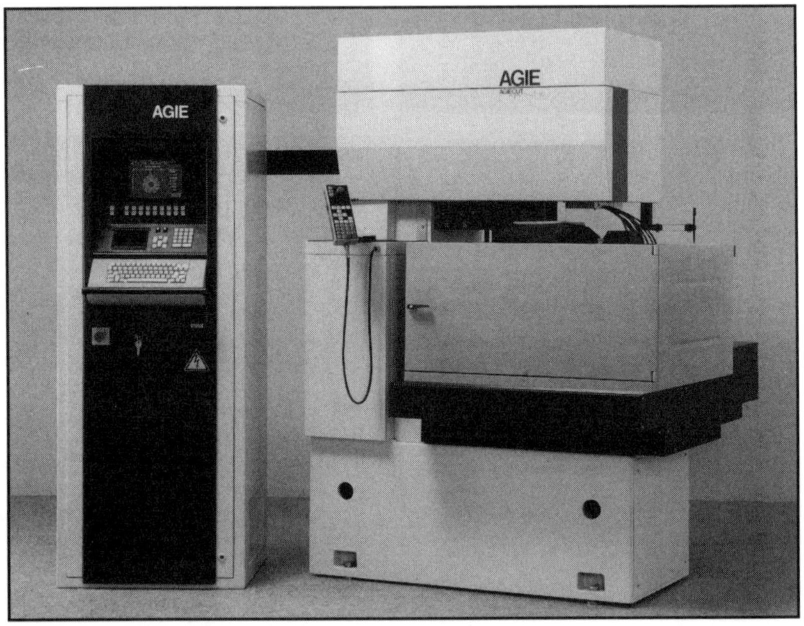

Courtesy Agie

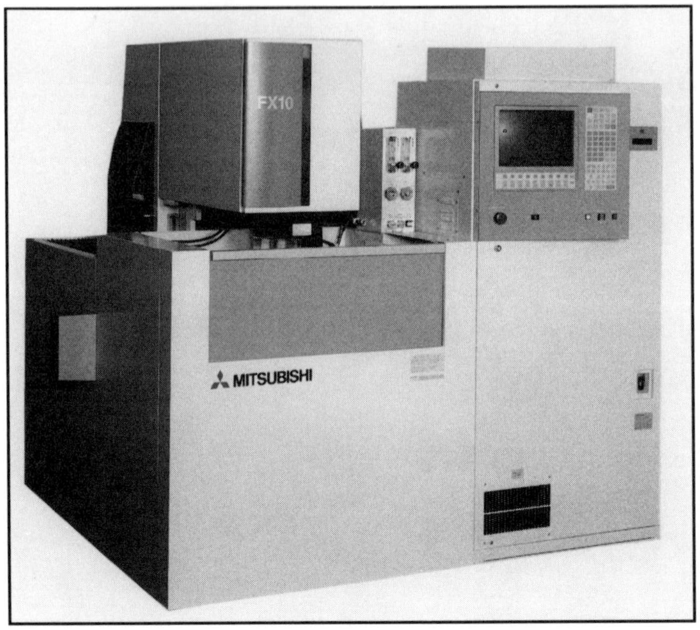

Courtesy Mitsubishi

Courtesy Sodick

High Precision Wire Electrical Discharge Machines

1

Fundamentals of Wire EDM

1 Fundamentals of Wire EDM

Revolutionizing Machining

Wire Electrical Discharge Machining (EDM), as shown in figure 1:1, has greatly altered the tooling and manufacturing industry, resulting in dramatic improvements in accuracy, quality, productivity, and earnings. In today's highly competitive world, every manufacturer must understand the many advantages of this latest technology in order to be competitive in the field.

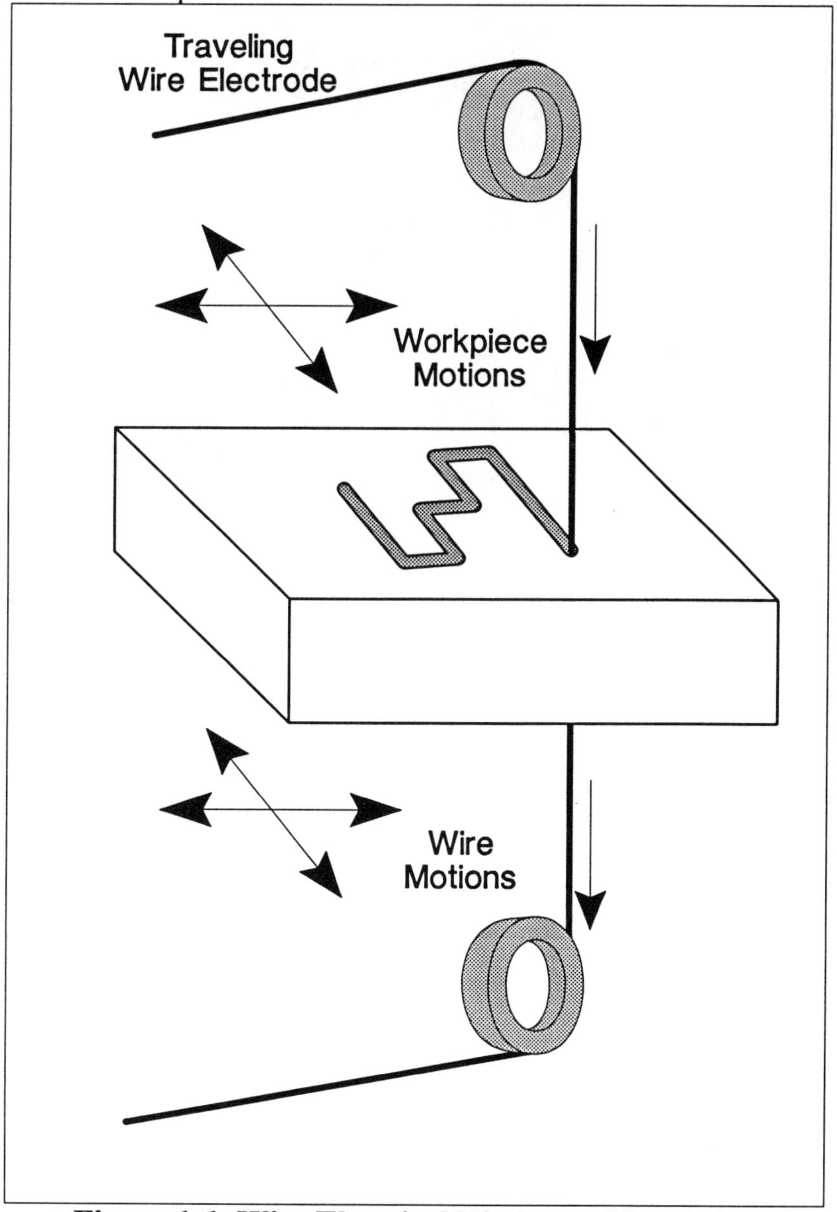

Figure 1:1 Wire Electrical Discharge Machining

In 1969, the Swiss firm Agie produced the world's first industrial path-controlled electrical discharge cutting machine. The first machines were extremely slow. In the early 70s a typical machine cut two square inches per hour, in the early 80s, six square inches per hour. Today typical machines cut over 20 square inches per hour, and some wire EDM manufacturers claim nearly 30 square inches per hour. Some of the machines are accurate up to ± .0002″ and producing surface finishes to 15 RMS and lower. Today's high-speed wire EDM machines, as illustrated in Figure 1:2, have become so efficient that they have revolutionized many machining procedures.

Figure 1:2 A Modern High Speed Wire EDM

A brochure published by the Society of Manufacturing Engineers states: "EDM can no longer be considered nontraditional machining. With improvements such as CNC, advanced power supplies, ability to handle larger work pieces, and advances in wire, EDM is now commonly used in everyday operations. These improvements are allowing the EDM user to obtain faster metal removal rates yet maintain super-smooth finishes with repetitive accuracy."

Senior Editor Frederick Mason, a regular reporter on EDM and machining, writes in the *American Machinist*, "Wire EDM has come out of the toolroom in the last half-dozen years and has become a technology for production machining across not only several industries but a wide range of part materials, sizes, tolerances, finishes, and prices as well."

Mason also notes, "Today wire EDM competes with conventional machining processes such as broaching, milling, and drilling, with conventional forming processes such as stamping and fine-blanking, as well as with other newer technologies such as laser-cutting... Some of the parts are what you would expect from EDM--expensive parts with very precise shapes in difficult-to-machine materials. But quite a few are surprisingly not in that mold at all. They are simple shapes in standard everyday steel and aluminum... So cost-per-part, JIT delivery, and quality, alone or together, make wire-EDM the process of choice."

One large wire EDM company reports that 65% of their cutting time is taken by production runs up to 30,000 pieces. On a job which would have required fine blank tooling and a 10 to 12-week wait to get the tooling, they finished the hardened, .062 inch thick stainless steel parts, burr-free and on time for their production schedule.

Whether cutting soft aluminum, hot rolled steel, super alloys, or tungsten carbide, manufacturers are discovering that these materials are often cheaper to produce on today's high-speed wire EDM machines than on conventional machines, with the added benefit of much greater quality. See Figure 1:3.

Figure 1:3 Wire EDM Skim Cutting

NEW DEMANDS BY DESIGN ENGINEERS

Design engineers are discovering the many advantages of wire EDM, and they are incorporating new designs in their drawings. Contract shops must understand wire EDM in order to properly quote these new designs.

Increasingly, drawings are calling for tighter tolerances, shapes that can only be efficiently performed with wire EDM, and more difficult to machine materials. With wire EDM these exotic steels often cut just as easily as mild steel. In fact, when wire EDM manufacturers select steel to demonstrate the efficiency of their machines, they choose hardened D2, a high chrome carbon tool steel.

TWO BASIC TYPES OF EDM

A. Plunge EDM

Also known as conventional, sink, or ram EDM, this method, as shown in Figure 1:4, generally uses a graphite electrode; its distinct advantage is that it can produce a blind-shaped form in a workpiece.

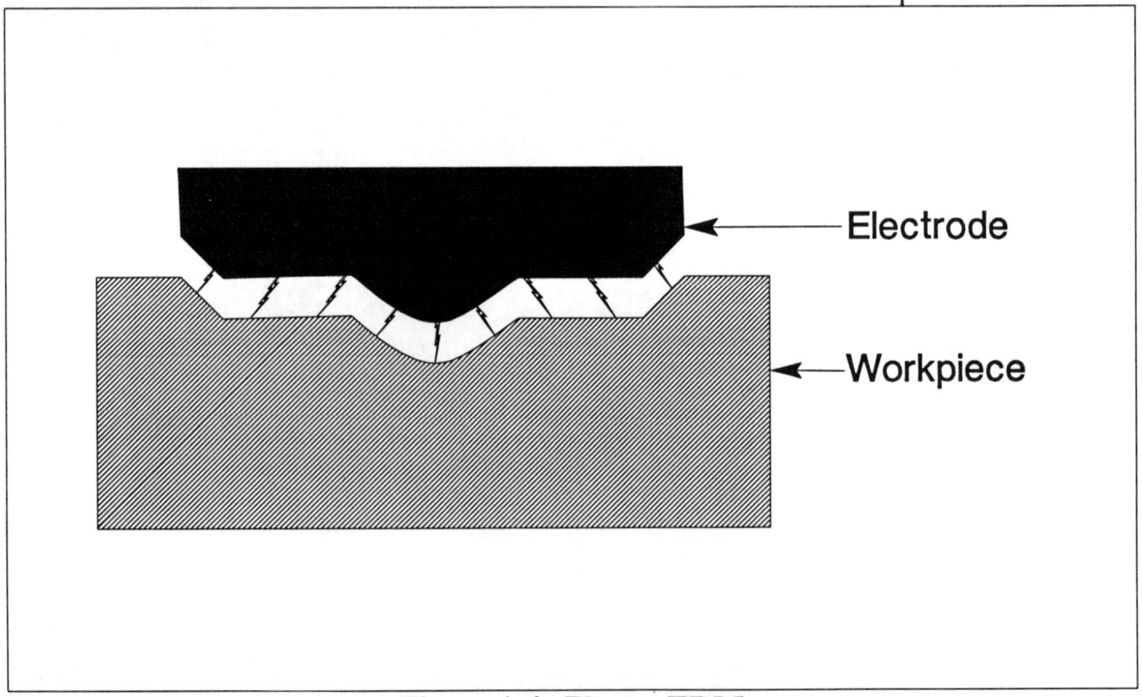

Figure 1:4 Plunge EDM

This EDM process uses a formed electrode to remove material.

B. Wire EDM

This process uses a traveling wire electrode that goes through the workpiece, and the wire is precisely monitored by a computer numerically controlled (CNC) system. See Figure 1:5.

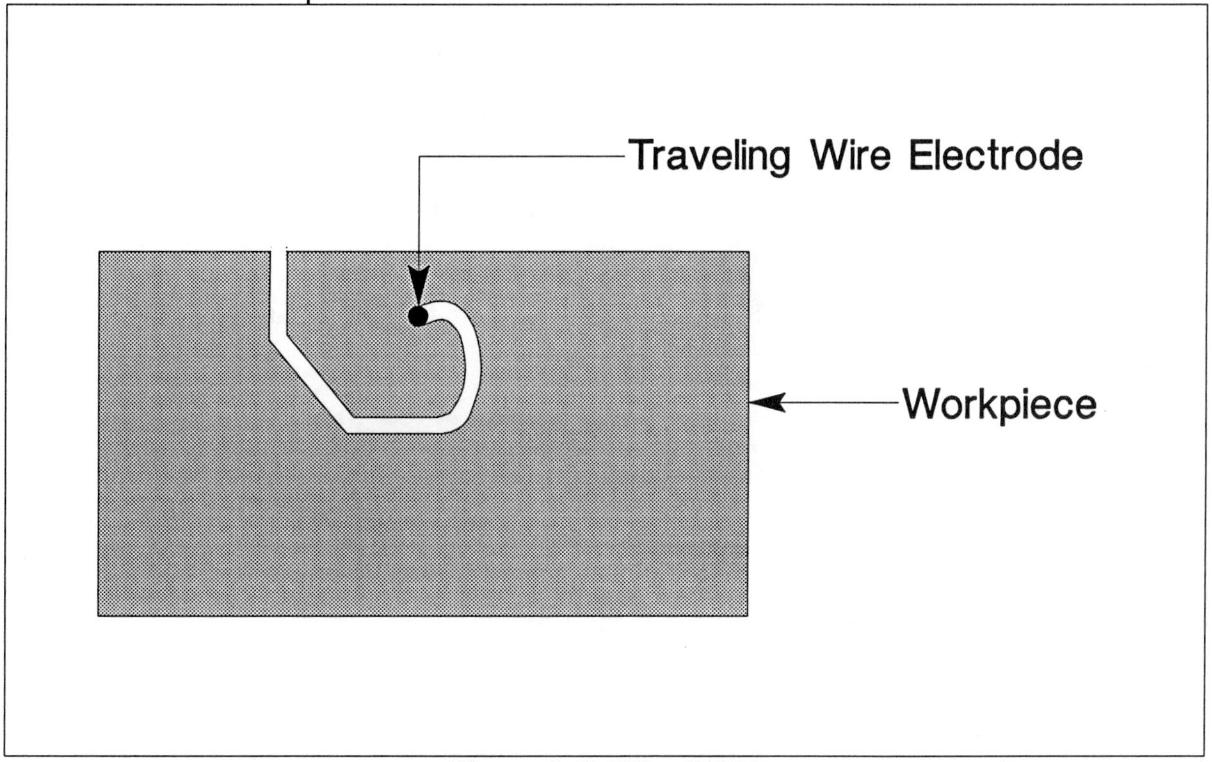

Figure 1:5 Wire EDM
This EDM process uses a computer controlled wire electrode to remove material.

HOW WIRE EDM WORKS

Wire EDM is a metal removal technique, just like any other machining tool; however, EDM uses electricity to remove metal by means of spark erosion.

Rapid D.C. electrical pulses are generated between the wire electrode and the workpiece with the aid of deionized water, called the dielectric. Between the wire and the workpiece there is a shield of deionized water. When sufficient voltage is applied and the fluid ionizes, a controlled spark precisely causes a small part of the workpiece to melt and vaporize. These pulses are repeated thousands of times per second. The pressurized cooling fluid, the dielectric, cools the vaporized metal and washes the resolidified eroded particles from the gap.

The dielectric fluid is composed of controlled ionized water. A filter is used to remove the suspended solids; a resin is used to control the electrical conductivity of the water. The dielectric fluid goes through a chiller in order to keep the liquid at a constant temperature to maintain machine accuracy. See Figure 1:6.

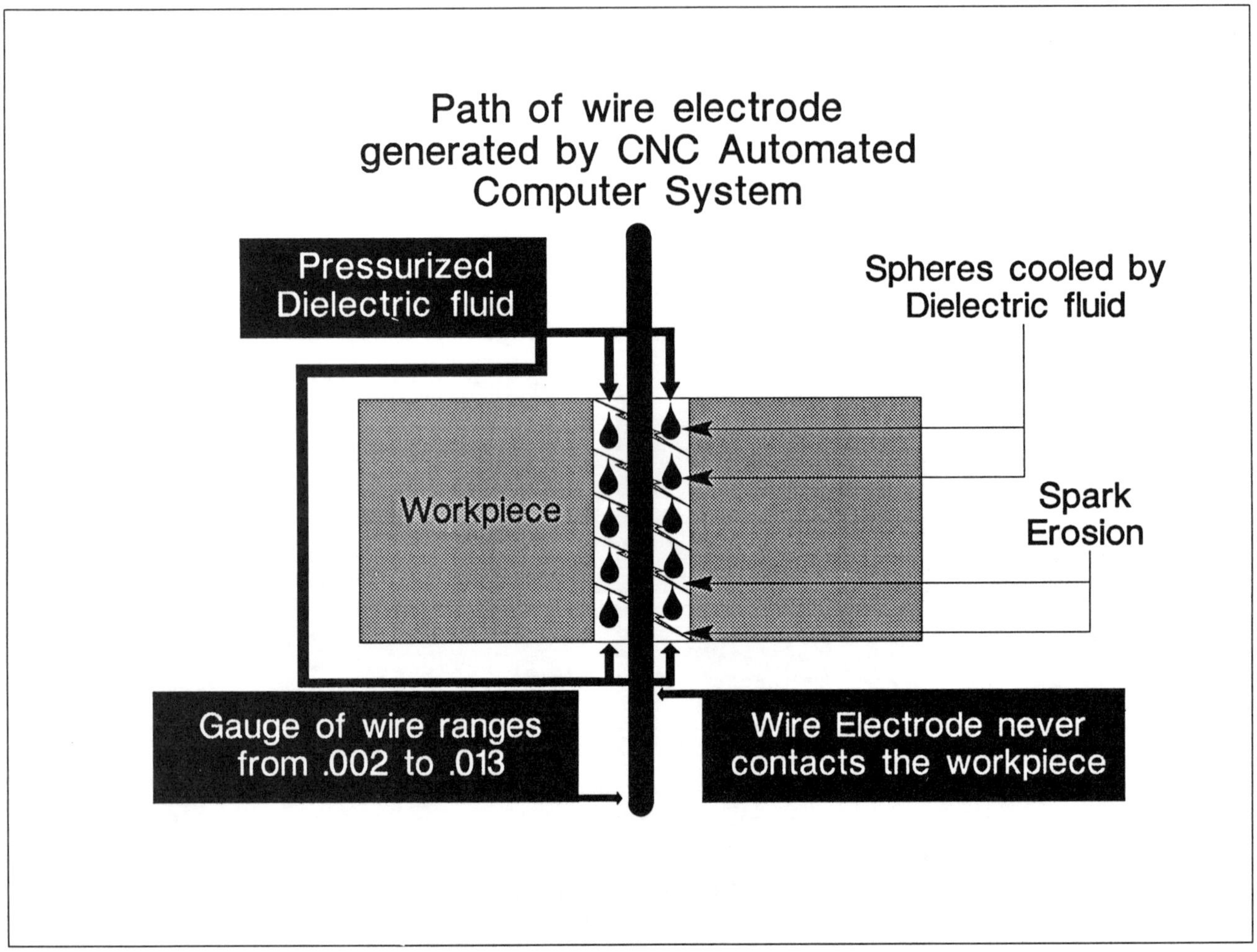

Figure 1:6 How Wire EDM Works
Precisely controlled sparks erode the metal using deionized water.
The pressurized fluid removes the eroded material.

A D.C. or A.C. servo system maintains a gap from .001 to .002 between the wire electrode and the workpiece. The servo mechanism prevents the wire electrode from contacting and from shorting out against the workpiece and advances the machine as it is cutting out the desired shape.

The wire electrode is usually a spool of brass, or brass and zinc wire, from .002 to .013 thick; sometimes molybdenum wire is used. New wire is constantly being used which accounts for the extreme accuracy of wire EDM.

THE STEP BY STEP EDM PROCESS

See Figures 1:7-10.

A. Power Supply Generates Volts and Amps

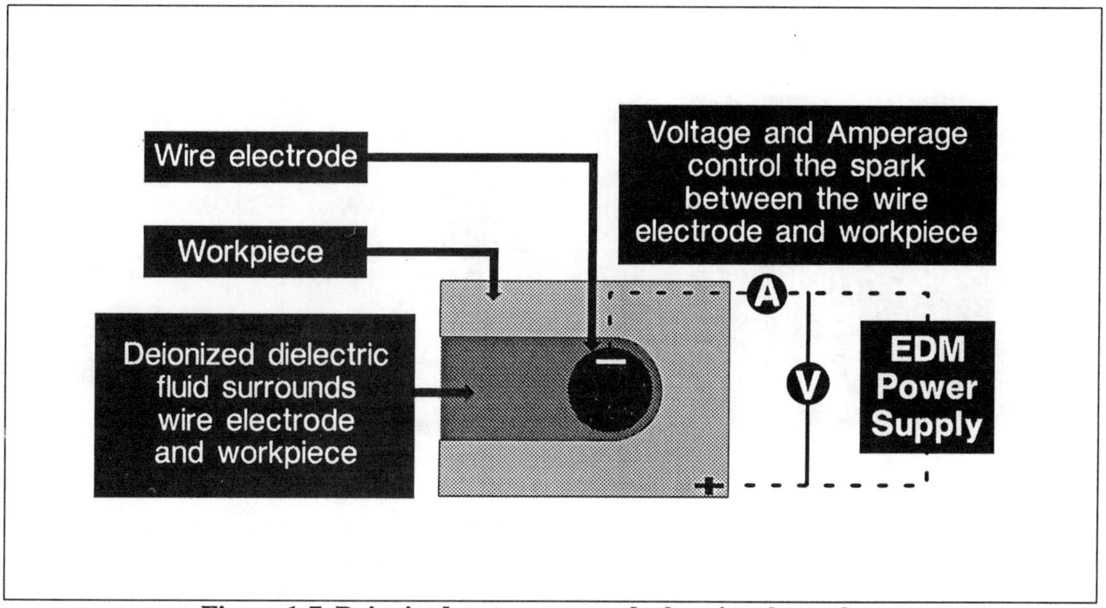

Figure 1:7 Deionized water surrounds the wire electrode as the power supply generates the volts and amps to produce the spark.

B. Controlled Spark Erodes Material During On Time

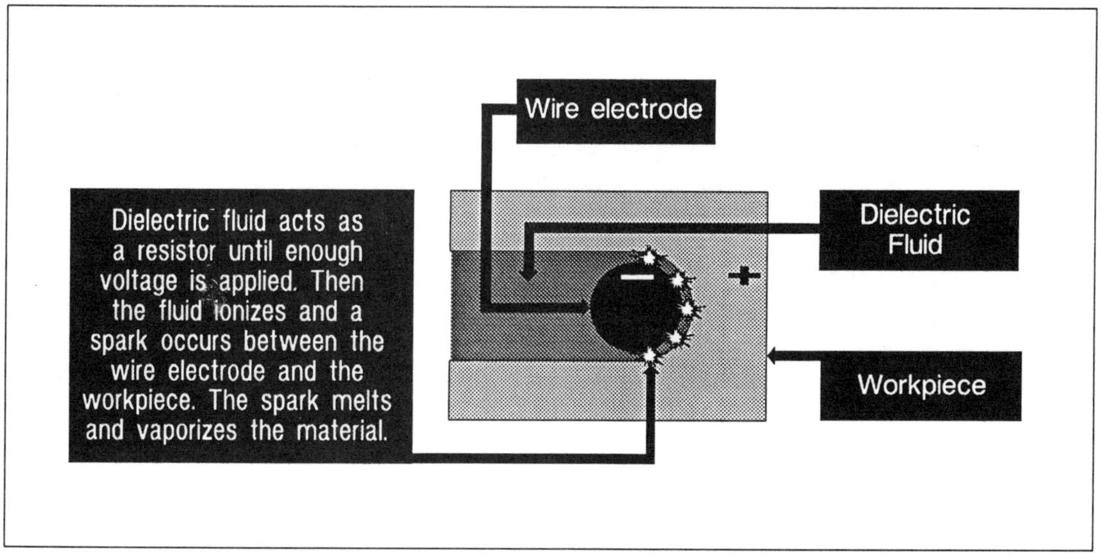

Figure 1:8 The generated spark precisely melts and vaporizes the material.

C. Off Time Allows Fluid to Remove Eroded Particles

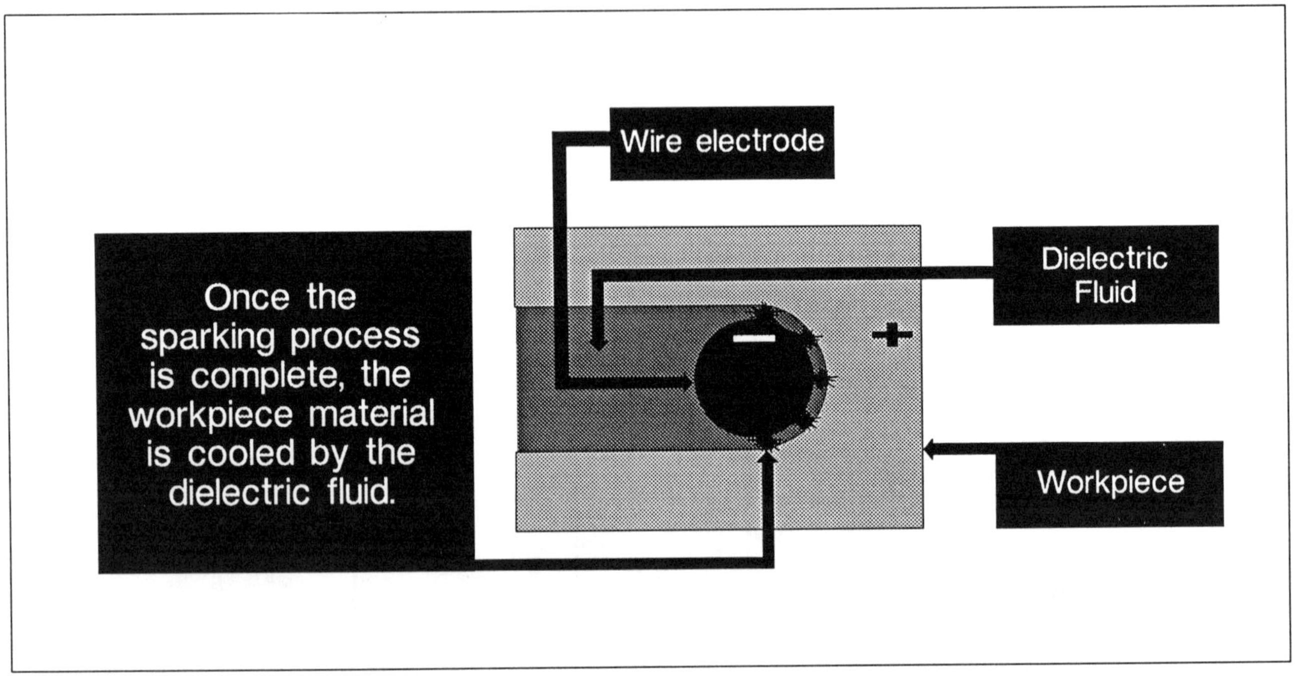

Figure 1:9 During the off cycle, the dielectric fluid immediately cools the material and flushes out the eroded particles.

D. Filter Removes Chips While Cycle is Repeated

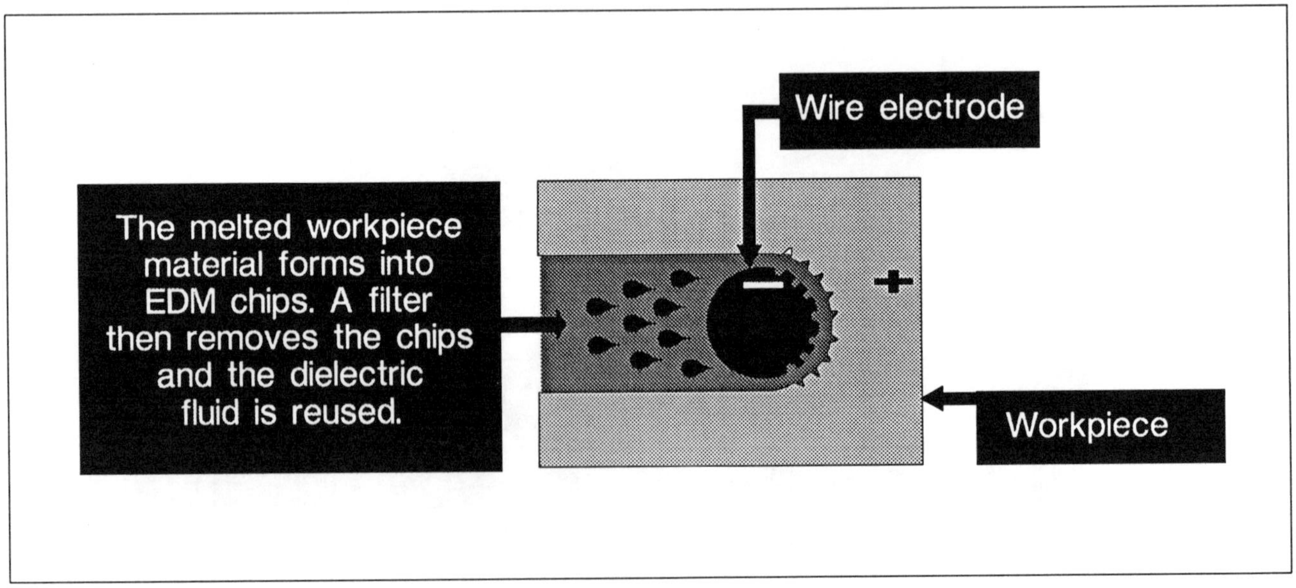

Figure 1:10 New Wire is constantly provided while the eroded particles are removed and separated by a filter system.

Super Precision Band Saw

To better understand the wire EDM process, one can visualize the wire EDM machine as a super precision band saw with accuracies up to +/-.0002. See Figure 1:11.

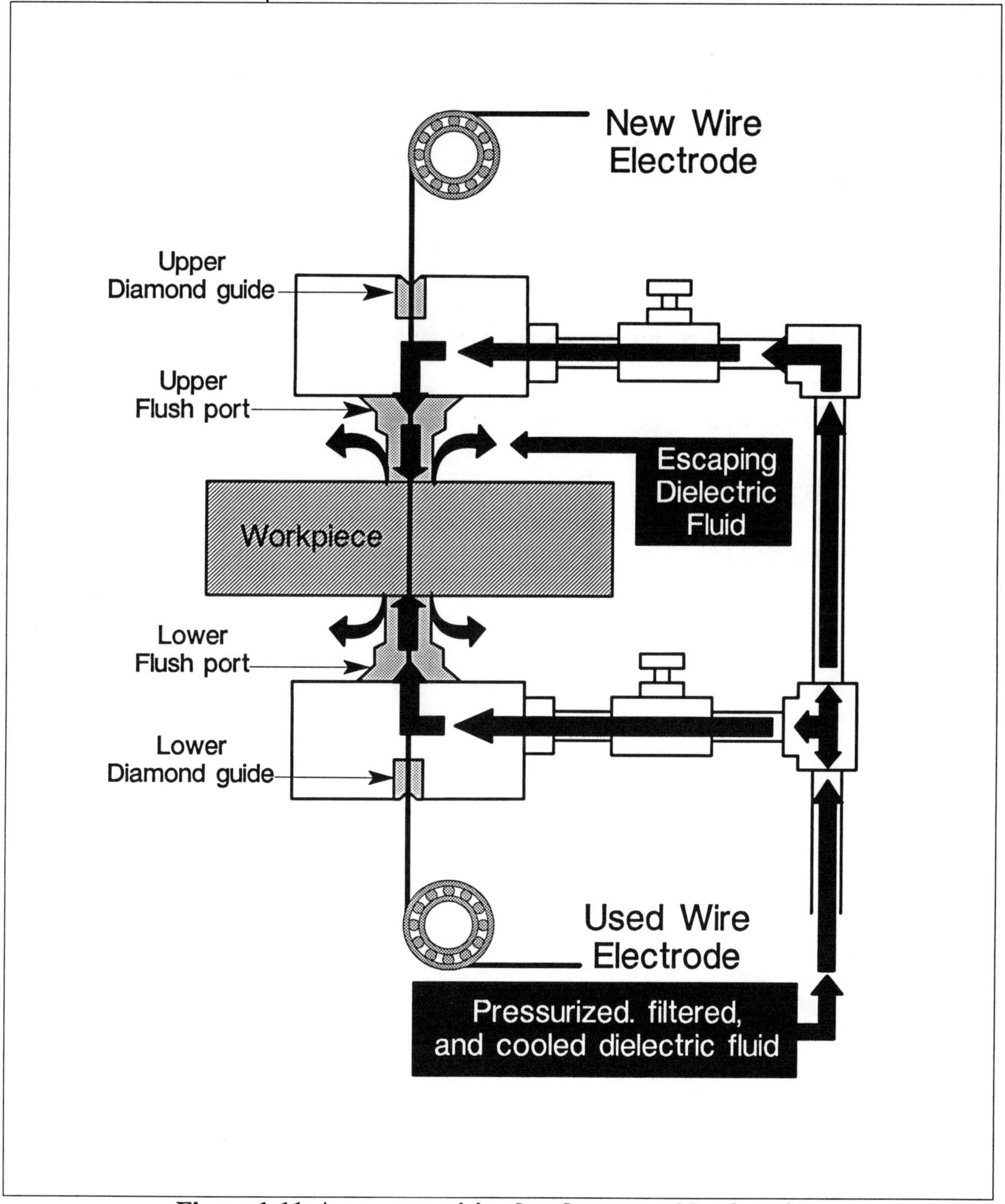

Figure 1:11 A super precision band saw capable of cutting hardened material to +/- .0002

THREE TYPES OF WIRE EDM

A. Two Axis

Only right angle cuts can be made as shown in Figure 1:12.

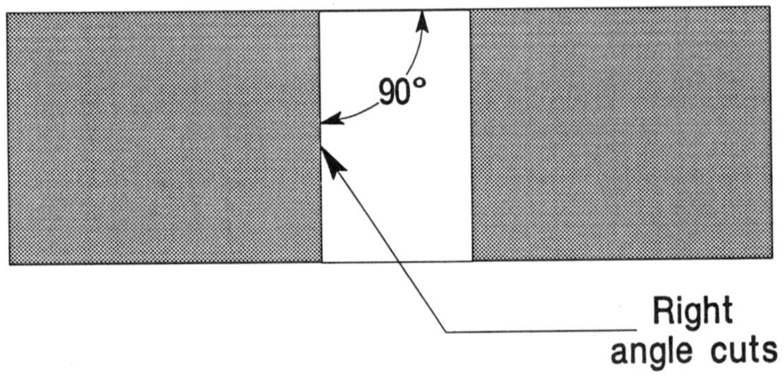

Figure 1:12 Two axis. For right angle cuts only.

B. Simultaneous Four Axis

These machines can produce a taper that is the same shape both on the top and on the bottom as pictured in Figure 1:13. They are also capable of going from a straight cut to a taper, particularly useful for making dies. A straight land can also be produced with a secondary cut.

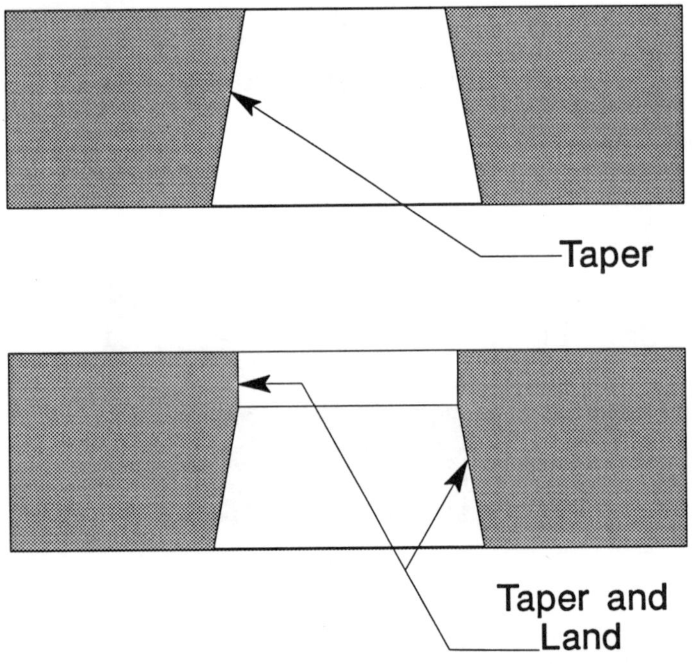

Figure 1:13 Simultaneous Four Axis
Taper and straight cuts can be produced.

C. Independent Four Axis

The top profile can be independent from the bottom profile as shown in Figure 1:14. This is particularly useful for extrusion molds and flow valves.

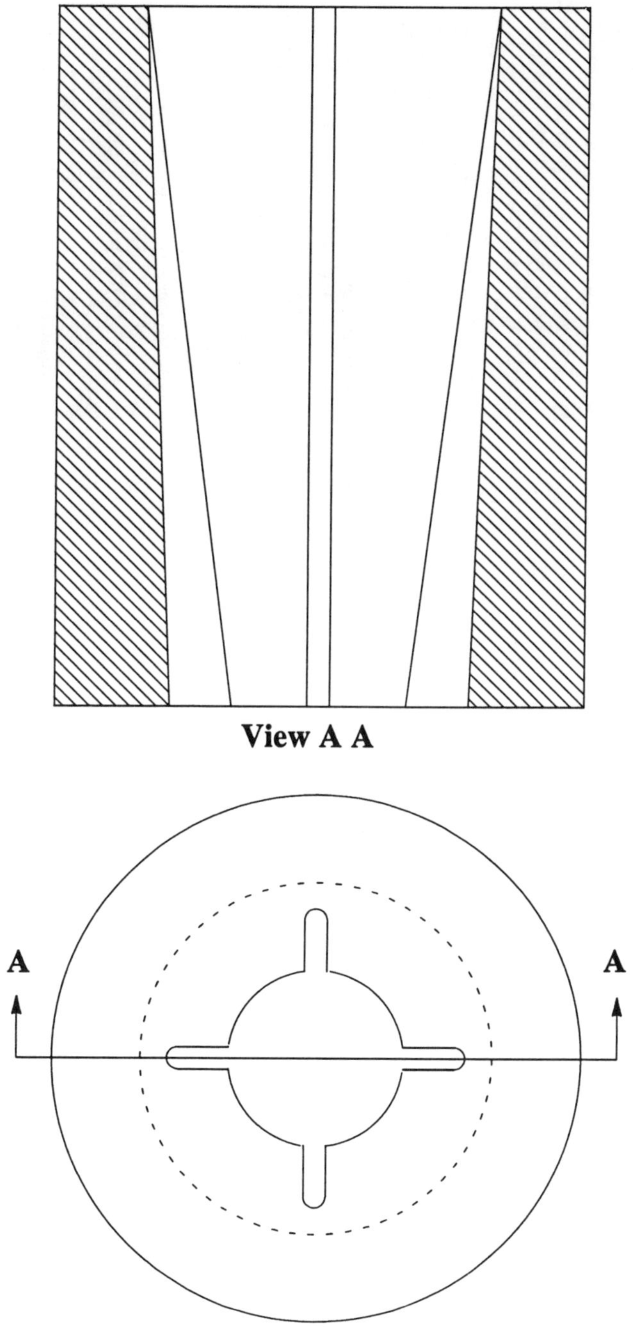

Figure 1:14 Independent Four Axis
Different shapes can be produced on the top of the workpiece from the bottom.

Parts shown in Figure 1:15 were produced with independent four axis.

Figure 1:15 Independent Four Axis Parts
(Part in lower right is the number one and two.)

A computer image of the numbers one and two are shown below in Figure 1:16.

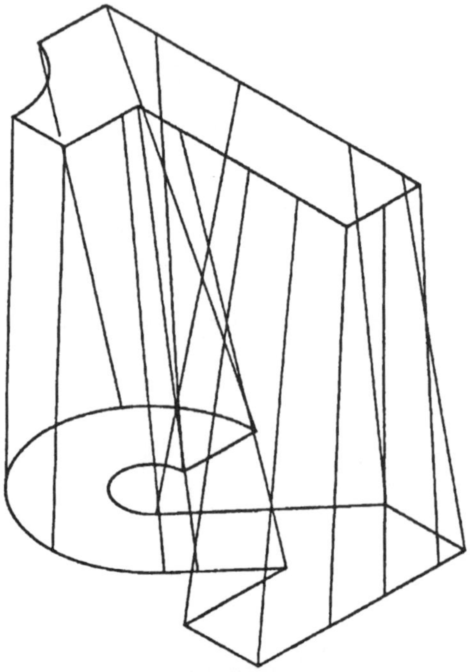

Figure 1:16

A picture of a cross and the Statue of Liberty is shown below in Figure 1:17.

Figure 1:17 Statue of Liberty Cross

A computer image of the Statue of Liberty Cross is shown below in Figure 1:18.

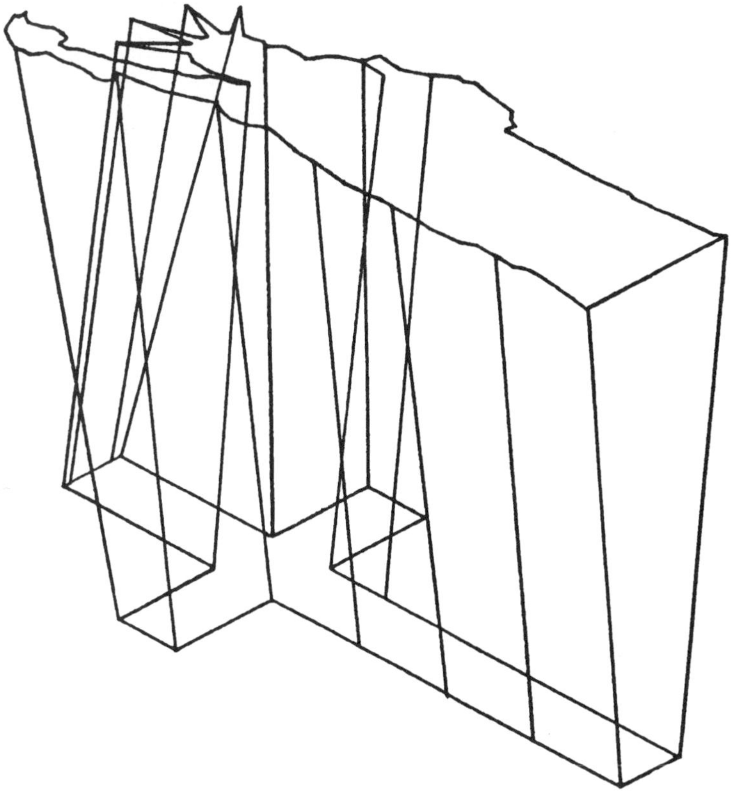

Figure 1:18

Understanding Independent Four Axis

Our customers have discovered many unique ways of using the capabilities of the independent four axis. In your design, picture holding a straight wire and moving the top and bottom independently. Virtually any conceivable shape can be created within the confines of the travel of the U and V axis.

Machines are now capable of cutting 15 3/4″ with independent angles up to 30 degrees. See Figure 1:19.

Courtesy Charmilles Technology

Figure 1:19
This machine is capable of cutting up to 15 3/4″ with independent angles up to 30 degrees.

2

Profiting With Wire EDM

2 Profiting With Wire EDM

Users of Wire EDM

Parts made with the wire EDM process are used in the field of medicine, in the chemical industry, in electronics, oil and gas production, the die and mold industry, in construction, the field of aeronautics, the automotive industry, manufacturing--virtually anytime metals are utilized.

Benefits Of Wire EDM

A. Production Runs

Because of the new generation of high-speed wire EDM machines, manufacturers are increasingly discovering that many parts can be more economically produced with wire EDM rather than with conventional machining. See Figure 2:1. An additional benefit is that close tolerances can be held without additional cost.

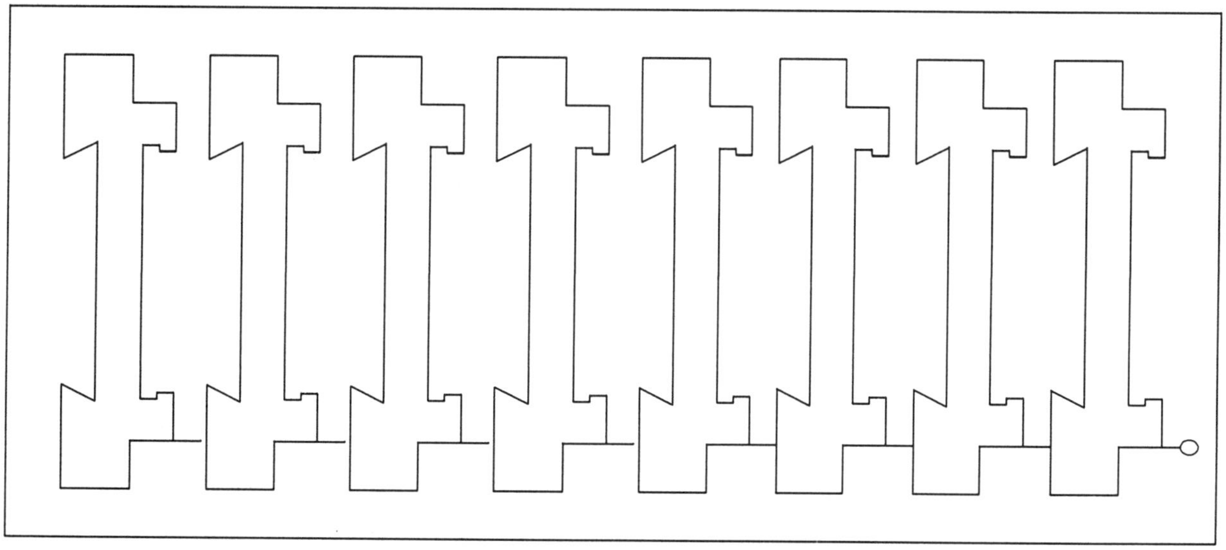

Figure 2:1 Production EDM
Today's high-speed wire EDM machines can produce many parts more economically than is possible with conventional machining.

B. Various Shapes and Sizes

With this new technology, any contour (Figure 2:2) and varying tapers can be precisely machined. Extremely thin sections can be made because the wire electrode never comes in contact with the material being cut. EDM is a non-contact, force-free, metal-removing process which eliminates mechanical distortion.

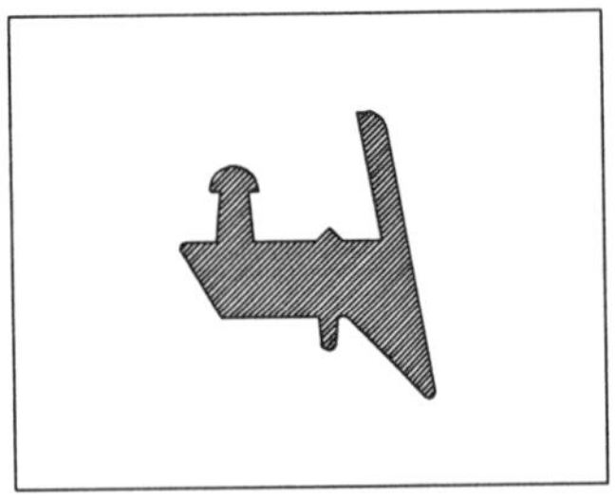

Figure 2:2 Various Shapes and Sizes
The most intricate shapes can be precisely machined.

C. Accuracy and Finishes

The wire path is controlled by a CNC computer-generated program with part accuracies up to +/- .0002 and on some machines achieves surface finishes below 15RMS. Dowel holes can be produced by wire EDM to be either press or slip fit.

D. Eliminates Extra Machining Processes

The extremely fine finish from the standard wire EDM process often eliminates the need for grinding or other finishing procedures. When wire EDM is being used, do not hesitate to add small radii to eliminate a secondary operation of deburring the edges (Figure 2:3). This does not effect the cost.

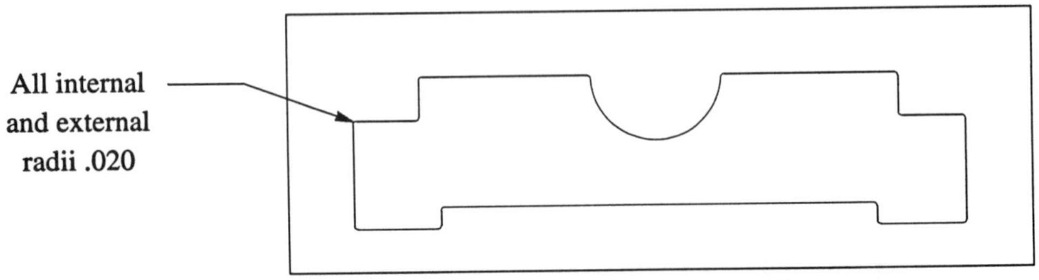

Figure 2:3 Eliminates Extra Machining Processes

E. Prototypes Economically Produced

Before making a part with expensive tooling, various prototypes can be cheaply produced as in Figure 2:4. Any needed changes can then be easily made on the computer, and a new prototype made.

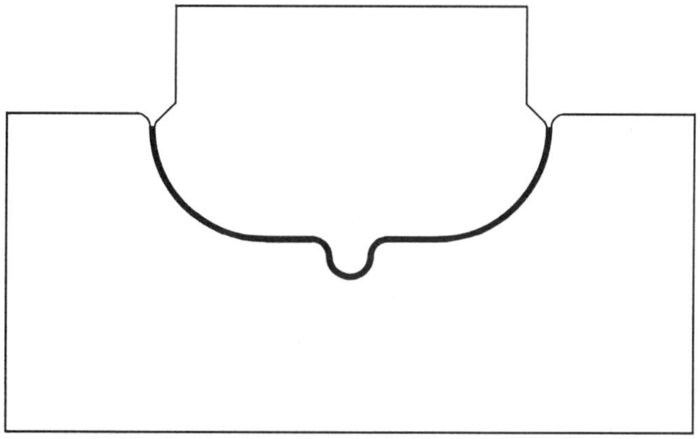

Figure 2:4 Experimental Blanks Can be Economically Produced
Prototypes for form dies can be easily made and changed before hard tooling.

F. Burr Free and Perfectly Straight Machining

Materials cut with wire are totally burr free, and the edges are perfectly straight. See Figure 2:5.

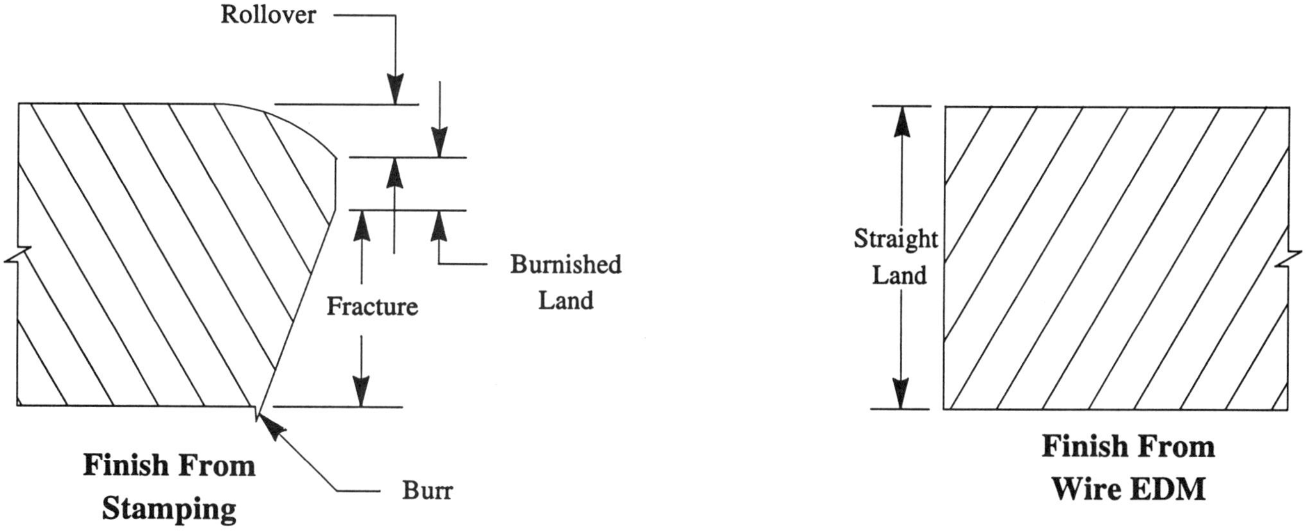

Figure 2:5 Wire EDM Parts are Straight and Burr Free

G. Damaged Parts Can Be Repaired with Inserts

When a die, mold, or machine part is damaged, it can be repaired with an insert rather than be remade. An insert can be made with a taper so that it becomes a forced fit. See examples in Figures 2:6,7.

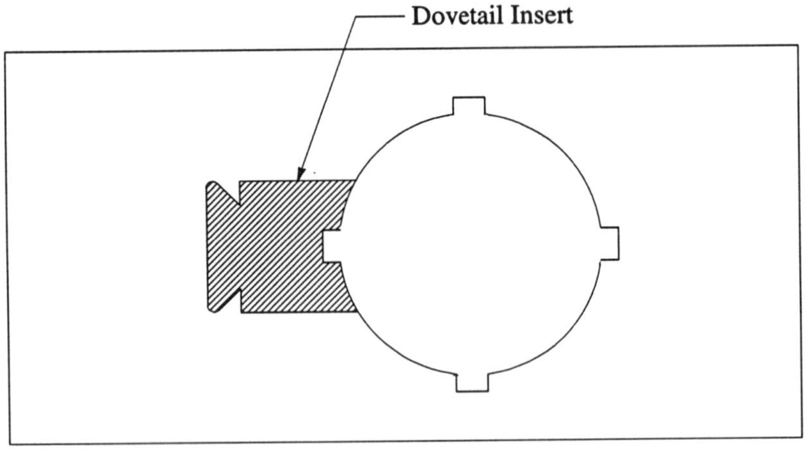

Figure 2:6 Damaged Die Repaired with Insert
Dovetail can be pressed fit or held with screw.

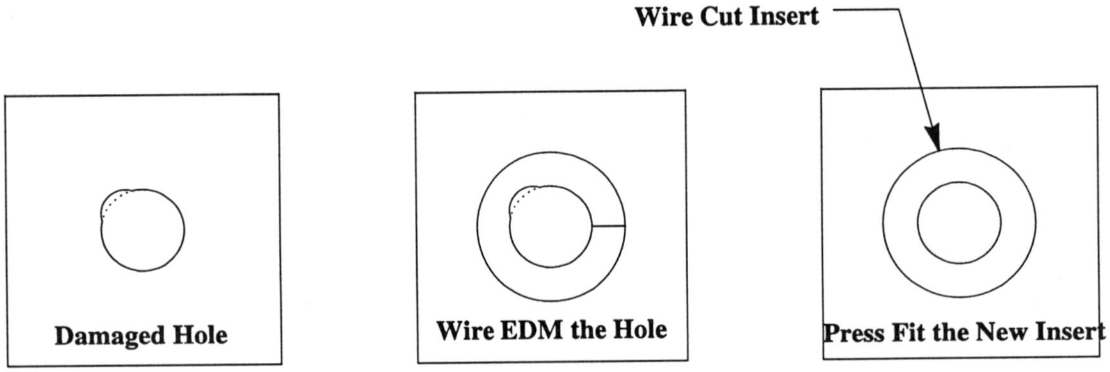

Figure 2:7 Repairing a Damaged Hardened Hole

H. Less Need for Skilled Craftsmen

Because wire EDM often eliminates extreme precision and time consuming machining processes, the need for skilled craftsmen is reduced, freeing those professionals for areas of work for which their skills are needed.

I. Material Hardness Not a Factor

Workpiece hardness does not effect the ability of cutting with wire EDM. In fact, hardened D2 cuts anywhere from 15% to 40%

faster than cold roll steel. The advantage of cutting in the hardened state is that it eliminates the risk of distortion created by heat treating.

The EDM process introduces very little heat into the material, and the small amount of heat generated is quickly removed by the dielectric fluid. Stamping dies in the hardened condition are consistently made from various tool steels with no negative results.

Figure 2:8 Computers Used To Program Path of Wire

J. Computer Calculations

Computers are used (Figure 2:8) to program the path of wire; consequently, in many cases only basic math dimensions are needed. No need to give exact chord positions on blending radii. The computer will automatically calculate the blending points as shown in Figure 2:9.

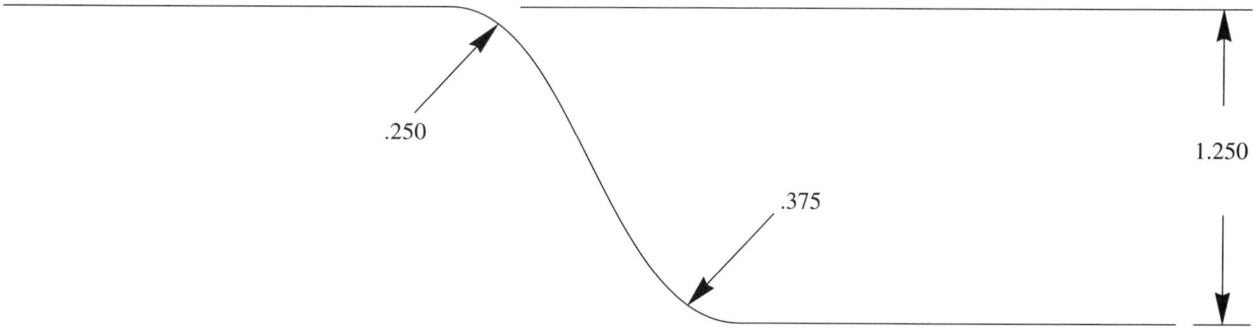

Figure 2:9 Computer Automatically Calculates Chord Position

K. Digitizing

Exact dimensions of a drawing are not necessary. With a digitizing plotter, a program can be made directly from a print as shown in Figure 2:10.

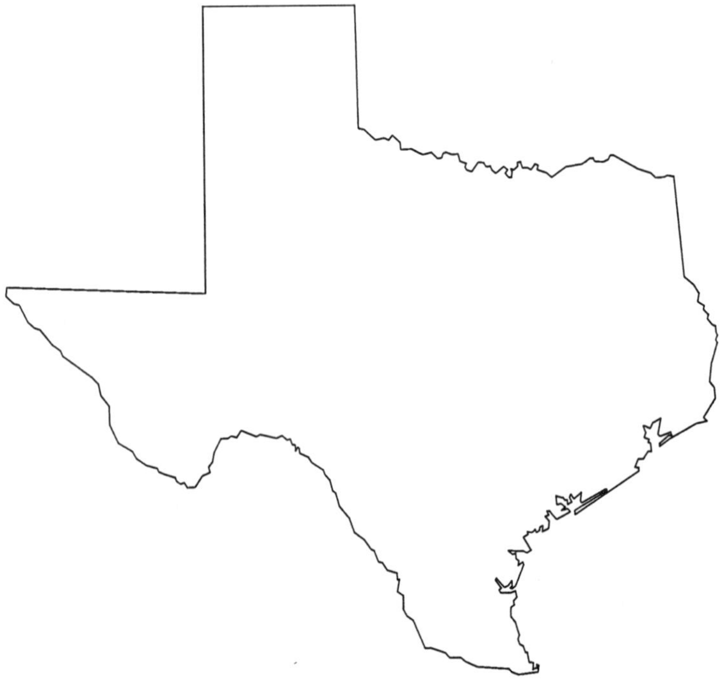

Figure 2:10 Example of a Digitized Drawing

L. Miniaturization

With extreme precision, wire EDM can machine thin webs (Figure 2:11), small holes, and close inside and outside radii with very fine micro finishes, because the wire never comes in contact with the metal.

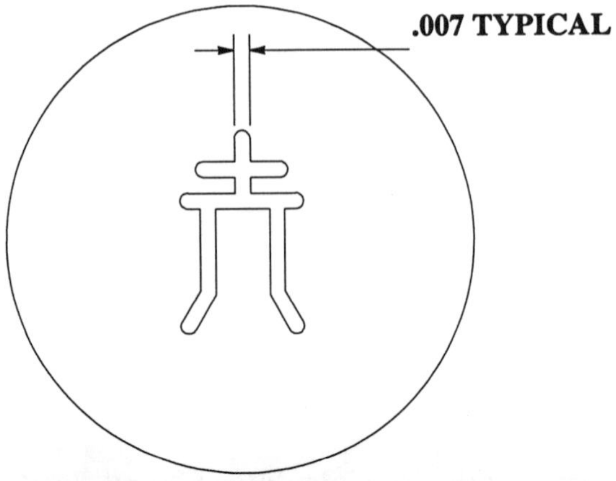

Figure 2:11 Miniaturization
Wire EDM can produce very thin webs.

M. Fabrication of Graphite Electrodes

Graphite electrodes for blind hole EDM can be produced with wire EDM. One of the great advantages for this is that each electrode produced with wire EDM will be identical.

The cost of producing graphite electrodes is largely determined by the cutting speed of the wire. There is a vast difference between the various grades of graphites. The Poco Angstrofine (EDM-AF5) graphite cuts nearly twice as fast as most of the other grades (EDM-1, EDM-3, EDM-100, or EDM 200).

N. Air Machining

Parts can be EDMed, as shown in Figure 2:12, even when flush ports are not directly against the part. Though air machining is a slower cutting process, many customers have found it economical to have parts wire EDMed.

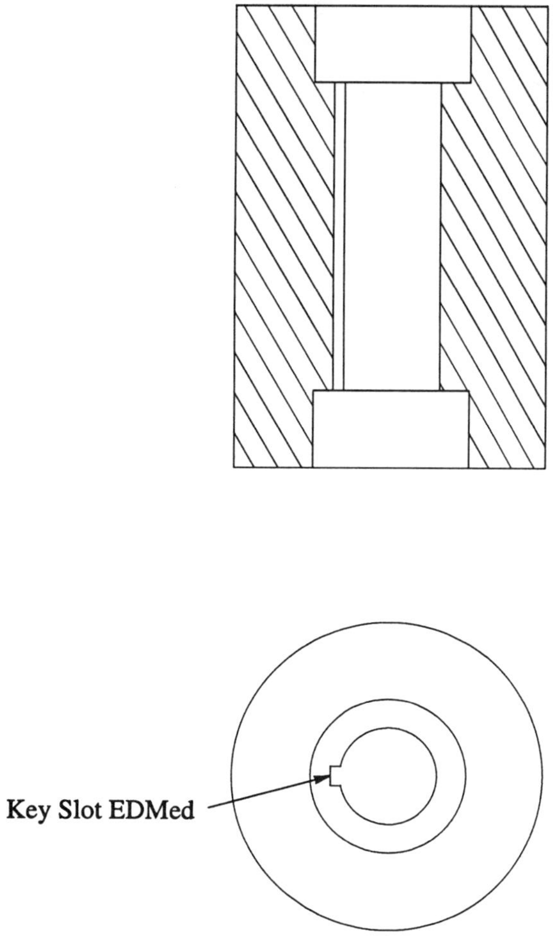

Figure 2:12 Air Machining

Keyway can be EDMed even though flush ports are away from part.

O. Reliable Repeatability

The constant reliability of wire EDM is one of the greater advantages of this process. Because the programs are computer generated and the electrode is constantly being fed from a spool (the wire electrode is used only once), the last part is identical to the first part. The cutter wear found in conventional machining does not exist. Additionally, tighter machining tolerances can be maintained, producing more accurate parts. See Figure 2:13

Figure 2:13 Quality Work
Display of wire EDM parts at Reliable EDM Corporation.

Parts for Wire EDM

A. Precision Gages and Templates

Instead of costly grinding procedures to make gages and templates, the program is created on a computer and then cut precisely with wire EDM as in Figure 2:14. Since gages and templates are often thin, making two or more at the same time does not add significantly to cost.

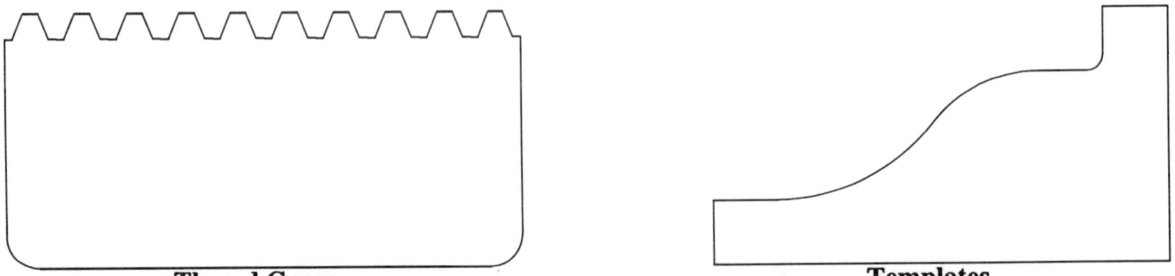

Figure 2:14 Precision Gages and Templates

B. Keyways vs. Broaching

Precision keyways can easily be cut with wire EDM, Figure 2:15, as can hex, splines and other shapes. No need to make a special broach; furthermore, the material can be in the hardened state.

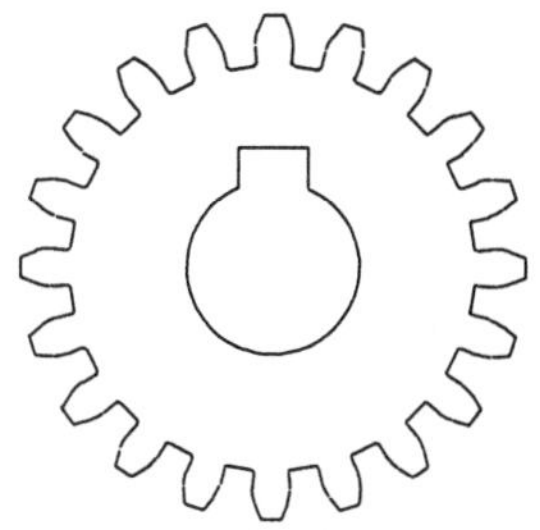

Figure 2:15 Precision Keyways

C. Shaft Slots

Rather than make a costly setup to machine a slot in a shaft, a simple setup can be made on a wire EDM machine. Additional to time and cost savings, no burrs result in the threaded area. See Figure 2:16

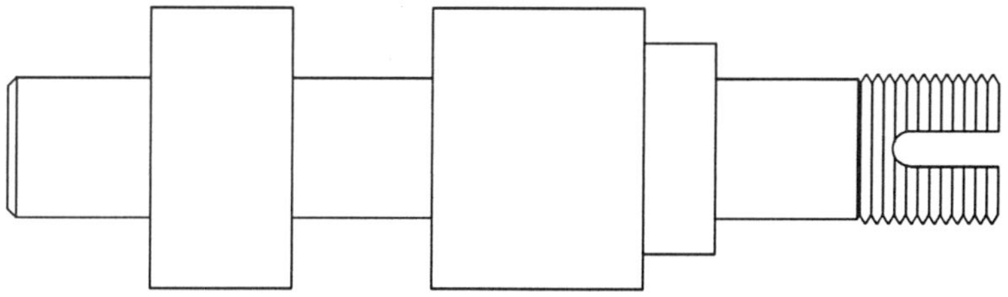

Figure 2:16 Burr Free Slot in Threaded Area

D. Collets

Conventional machining often distorts collets. If the collets are heat treated, they often distort even more. In contrast, wire EDM can cut collets (as shown in Figure 2:17) in the hardened condition and without cutting pressure.

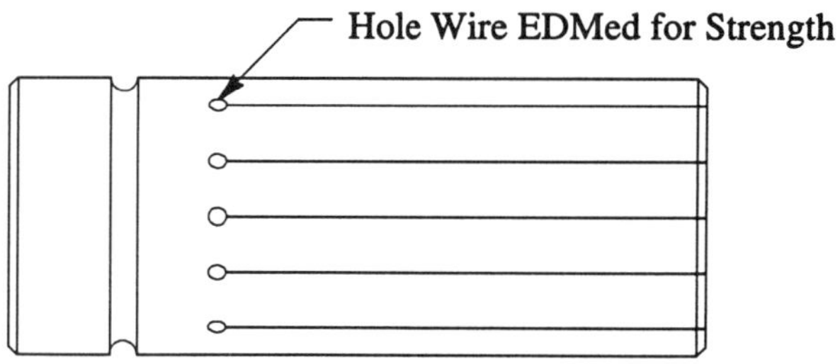

Figure 2:17 Collets

E. Parting Tubes and Shafts

Because of the small gap produced by wire EDM (a .012″ wire produces a .015″ gap), tubes, shafts, and bearing cages can be parted after machining is completed as pictured in Figure 2:18.

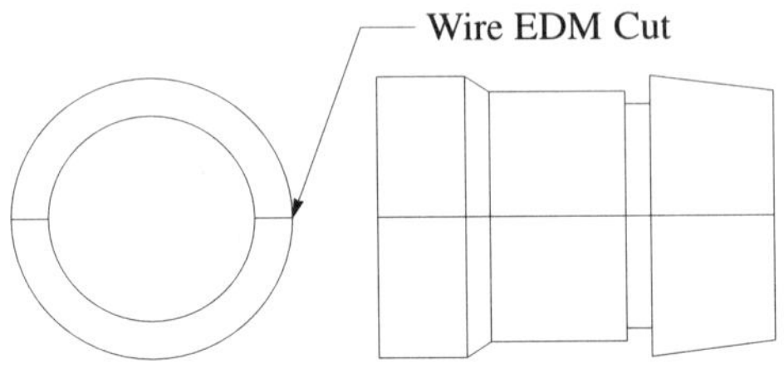

Figure 2:18 Splitting Tubes

F. Shaft Pockets

Any shaped pocket as shown in Figure 2:19 which goes through the shaft can be machined with wire EDM.

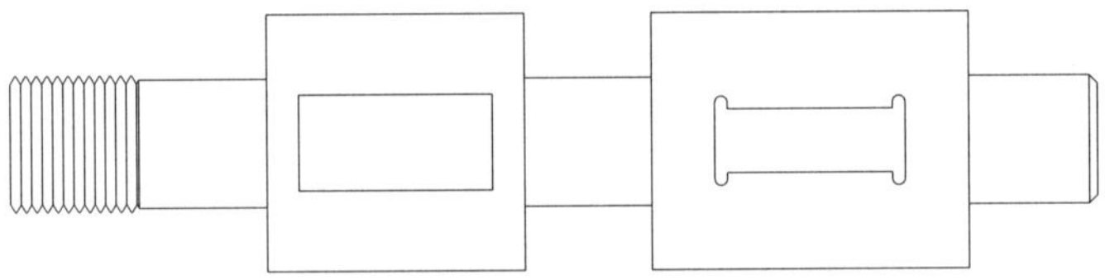

Figure 2:19 Shaft Cutouts

G. Punches and Dies From One Piece of Tool Steel

Dies no longer have to be made by the costly method of being sectioned and precision-ground. Now the most elaborate contours can be made in one solid piece, as shown in Figure 2:20, resulting in a stronger, non-breathing die at a fraction of the cost of a sectioned die. Chapter 10 provides detailed instruction on producing these long-life, inexpensive dies.

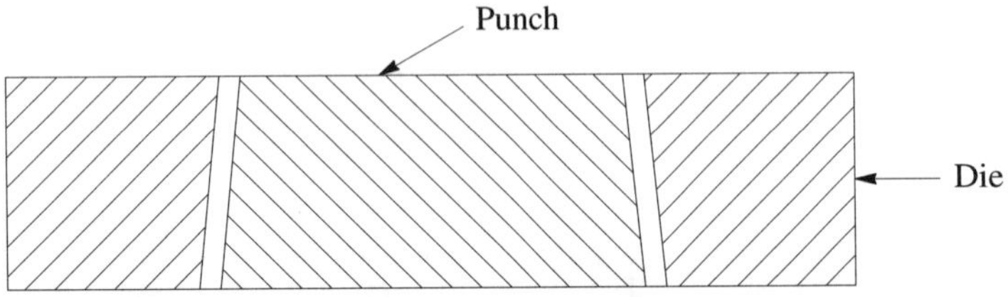

Figure 2:20 One Piece Punch and Die

H. Short Run Stampings

Instead of expensive tooling being produced, short run parts can often be much more economically produced with wire EDM. (Figure 2:21) Alterations are possible at practically no cost, while alterations with hard tooling are often very costly. All kinds of special shapes can be produced in various thicknesses.

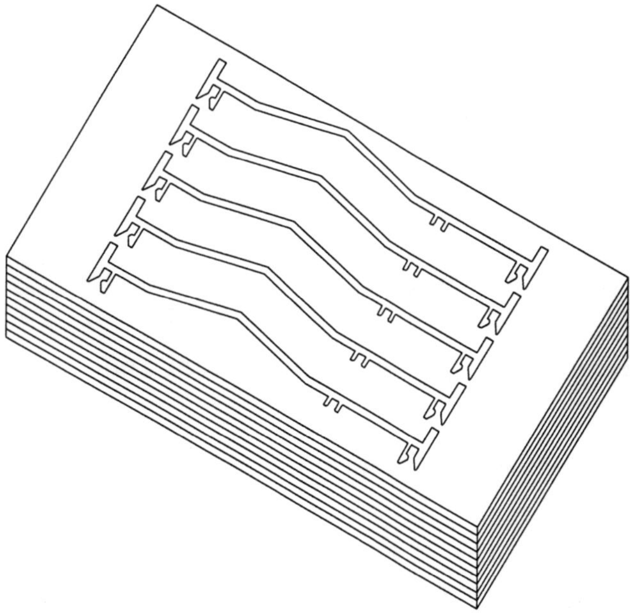

Figure 2:21 Stacked Material to Produce Intricate Parts

I. Molds

Elaborate extrusion molds, with or without taper as illustrated in Figure 2:22, can be economically produced.

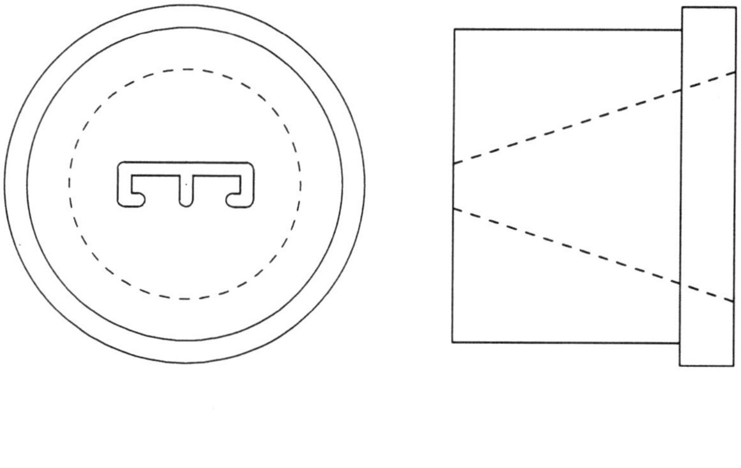

Figure 2:22 Tapered Extrusion Mold

J. Special and Production Tool Cutters

Special one-of-a-kind tooling, including carbide as shown in Figure 2:23, can be produced with various tapers with wire EDM. Production tool cutters should be the same to eliminate costly setups and checking procedures when changing cutters. Since wire EDM repeats accurately, production cutters made with this process will be identical.

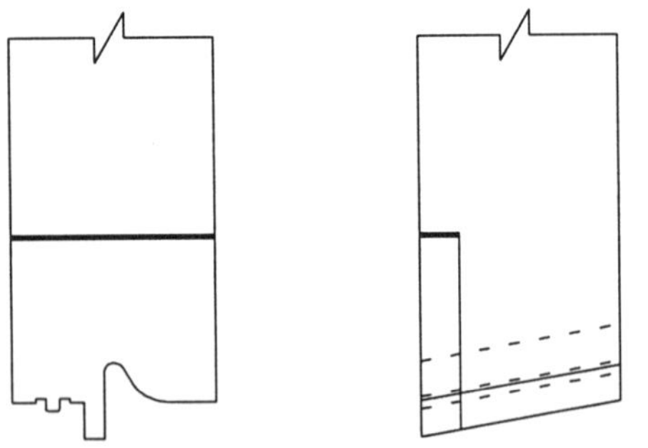

Figure 2:23 Special Carbide Form Tool

K. Difficult-to-Machine Shapes

Wire EDM has dramatically reduced costs for many manufactured parts. Machine shops are increasingly becoming aware of the benefits of EDM. Instead of using costly setups and complicated machining procedures, it is being discovered that wire EDM is often more cost effective. See Figure 2:24 for a difficult production machining operation that could be economically produced with wire EDM.

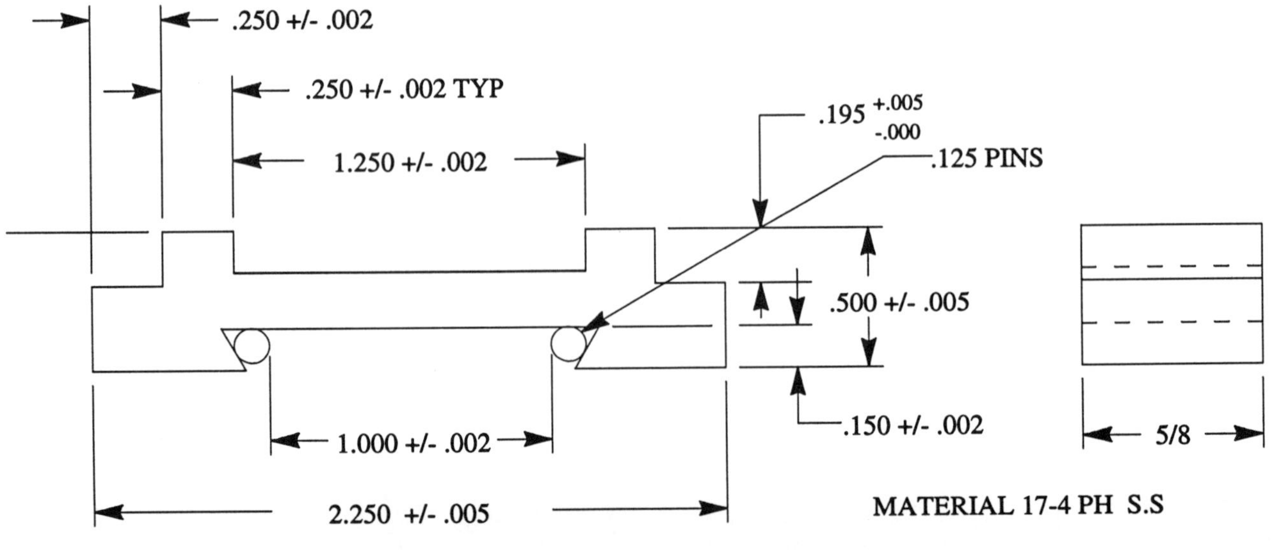

Figure 2:24 Difficult-to-Machine Shapes

L. Other Cost Reducing Parts

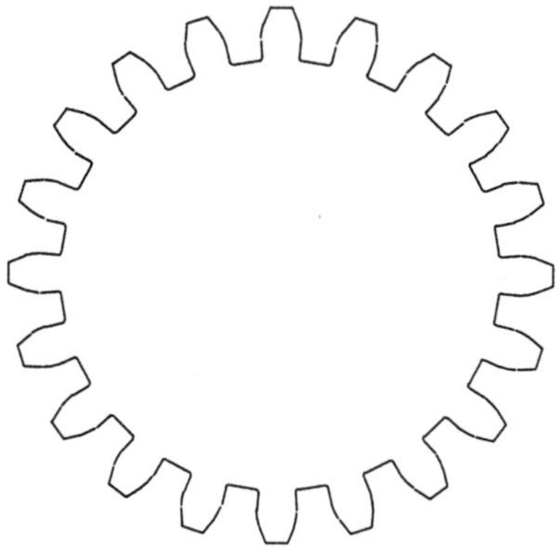

Figure 2:25 Gears

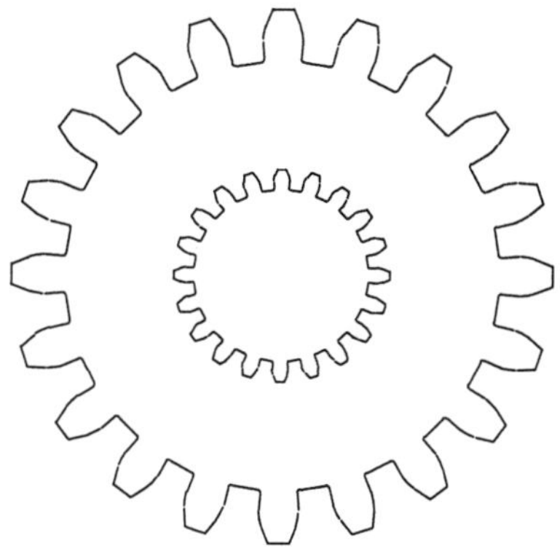

Figure 2:26 Internal Splines

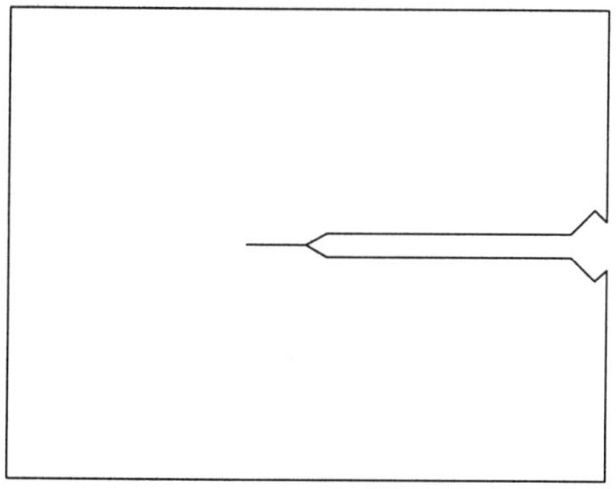

Figure 2:27 Notch Test Specimens

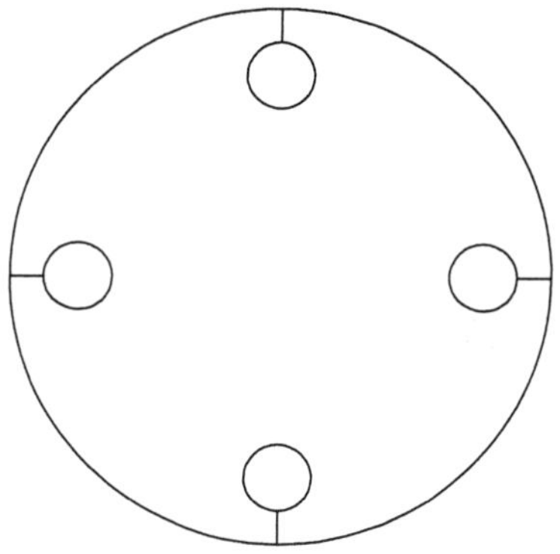

Figure 2:28 Round Test Specimens

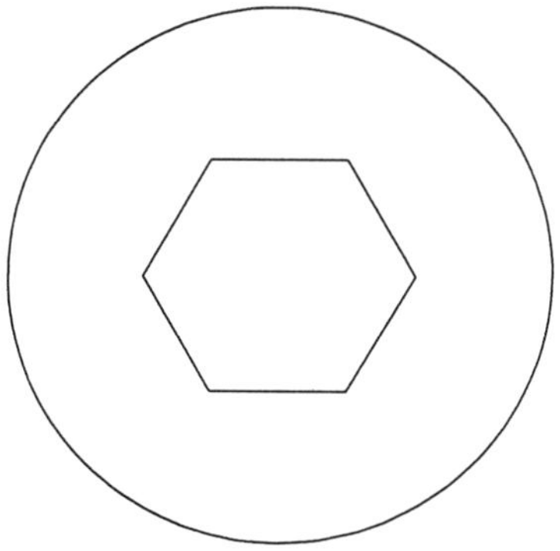

Figure 2:29 Hexes

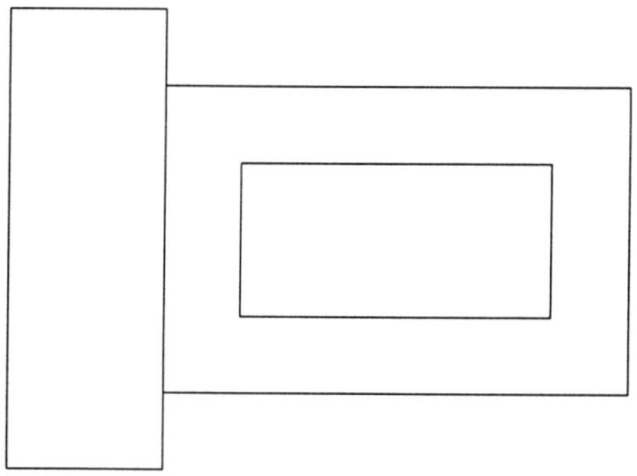

Figure 2:30 Rectangular Shapes

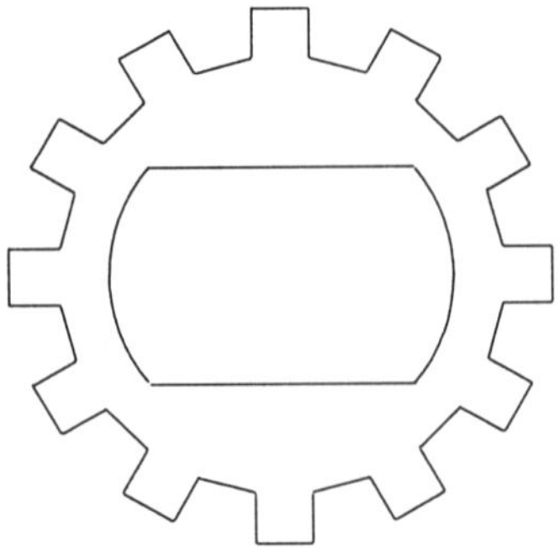

Figure 2:31 Special Cutters

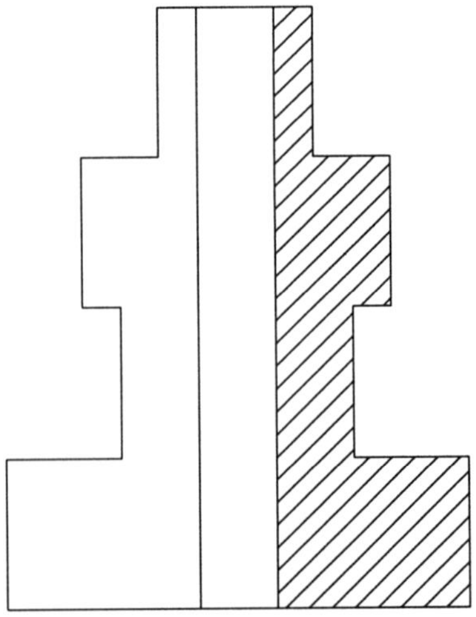

Figure 2:32 Special Shapes

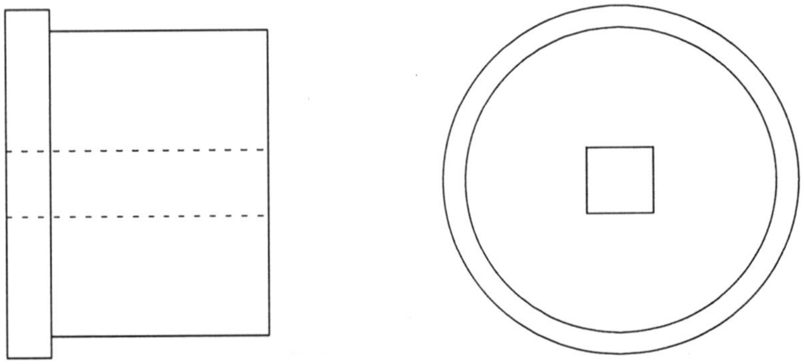

Figure 2:33 Carbide Nozzles

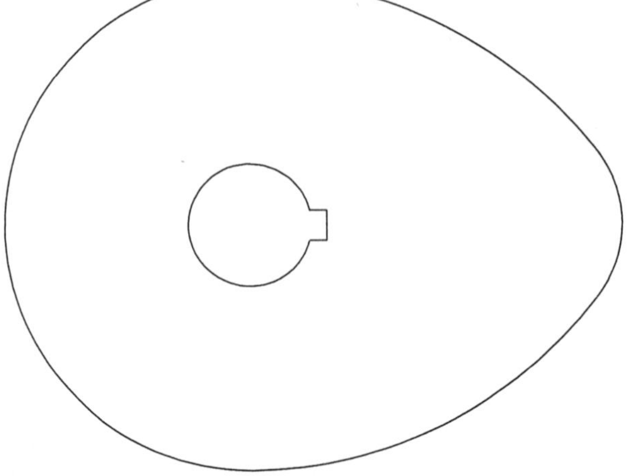

Figure 2:34 Cams

Advantages of Wire EDM vs. Plunge EDM

A. Producing Costly Electrodes

Plunge EDM is very effective for blind cavities; however, where shapes are throughout the piece, wire EDM is practically always more cost effective. The reasons are:

1. Plunge EDM as shown in Figure 2:35 needs a shaped electrode. Wire EDM uses wire from a spool.

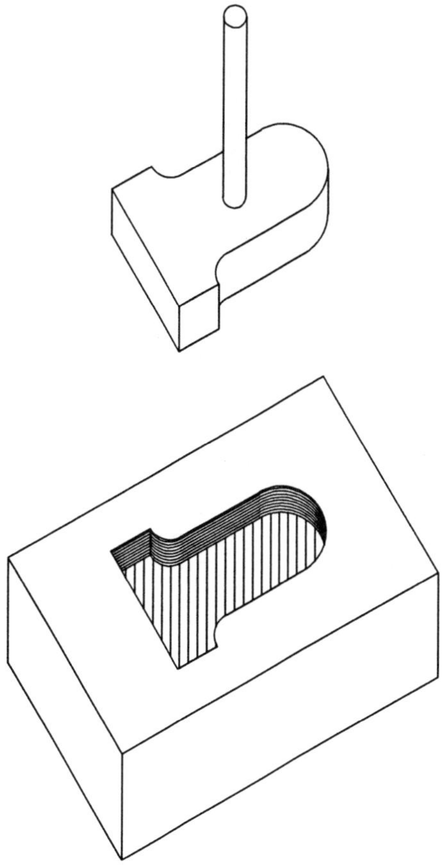

Figure 2:35 Plunge EDM
In plunge EDM the electrode must be machined.

2. Shaped electrodes are often costly to produce. Wire EDM never uses a custom shaped electrode. A computer controls the wire movement.

3. Electrodes in plunge EDM wear. Sometimes roughing and finishing electrodes are needed. With wire EDM, new wire is constantly being fed, so electrode wear is not a major problem. In addition, more accurate work is achieved.

B. Designing Cost Effective Parts.

Design engineers can save money by redesigning parts for wire EDM instead of plunge as demonstrated in Figure 2:36.

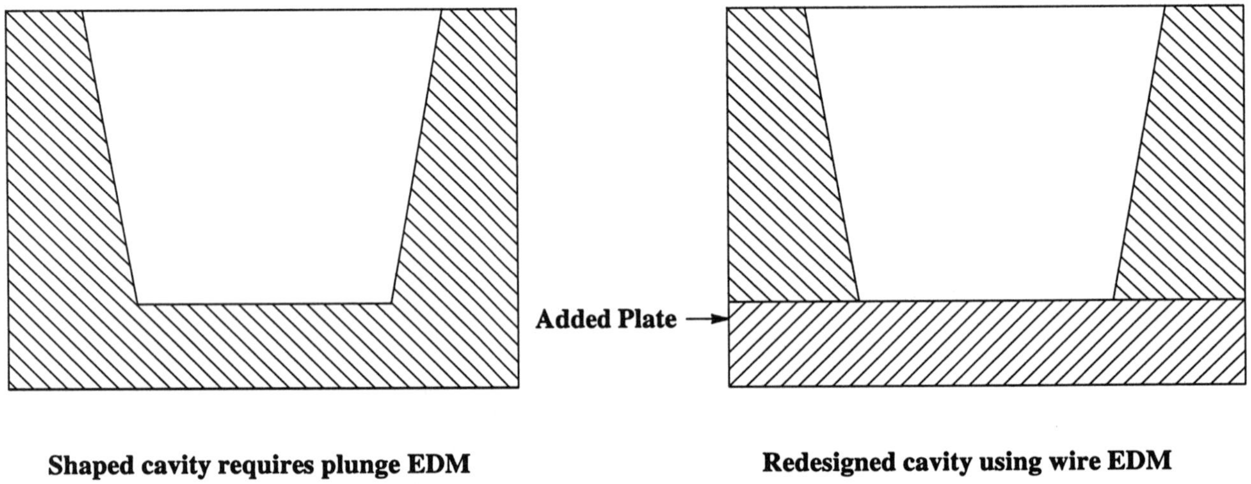

Figure 2:36 Redesigning Parts to Reduce Costs

Advantages of Wire EDM vs. Laser

Lasers are extremely efficient cutting machines. Lasers are thermal heating devices like wire EDM, but lasers get their heat from the amplification of light, whereas wire EDM receives its heat from electricity. Wire EDM needs material that is electrically conductive; lasers can be used in cutting both conductive and non-conductive materials. Lasers, however, have difficulty with reflective materials.

A. Better Surface Finish and Accuracy

Wire EDM gives a better surface finish and is much more accurate. When cutting plates, laser edges tend to be quite rough. Generally, 1500 watt CO^2 lasers are not effective over 1/2" thick in steel and over 1/4" in aluminum. The new high-speed machines have made wire EDM much more competitive with lasers. Though lasers have advantages, when accuracy and a fine surface finish are desired, the preferred method is wire EDM.

B. Cutting Shim Stock Absolutely Burr Free

For most sheet metal parts lasers are more cost efficient. However, when thin material needs to be cut without many holes wire EDM can be significantly cheaper and with an edge that is totally burr free. For example: Cut 500 parts from .005 shim stock. With wire EDM, shim stock is cut and sandwiched between two 1/4 inch plates of steel as shown in Figure 2:37.

Multiple parts can be cut as shown in Figure 2:38.

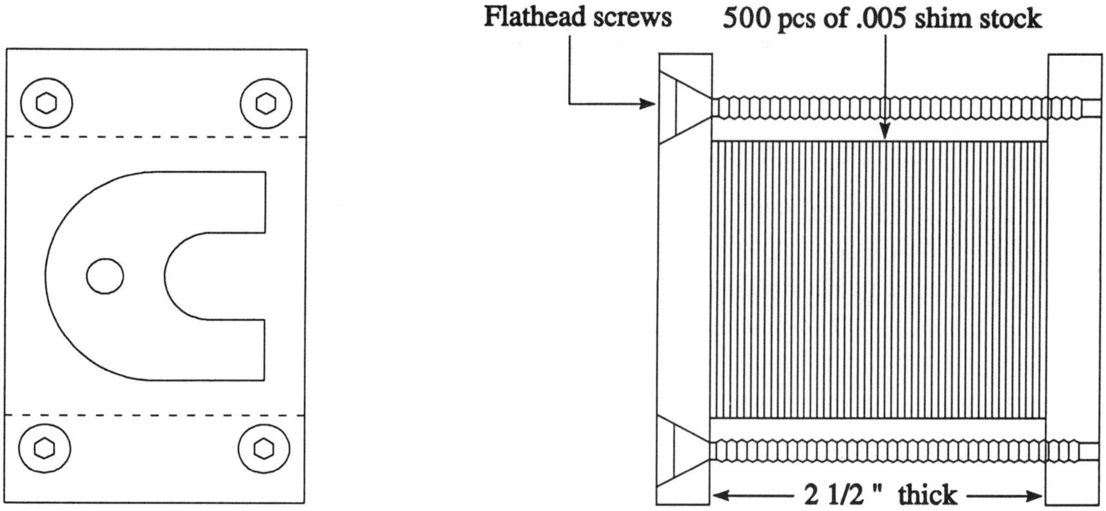

Figure 2:37 EDMing Shim Stock Burr Free

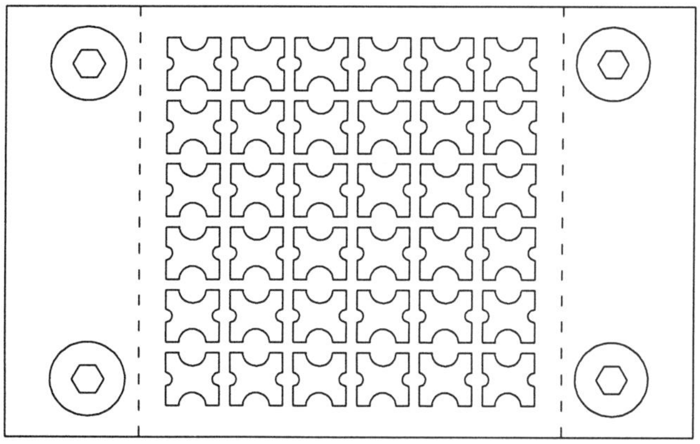

Figure 2:38 Cutting Multiple Shims

Materials That Can be EDMed

Wire EDM can cut any electrically conductive hard or soft substance. Listed below are some of the materials that can be wire EDMed.

Machining Costs

Inconel	Aluminum	Vasconal 300
Tool Steels:01, A2, D2, S7	Aluminum Bronze	PCD Diamond
Carbide	Copper	Nitronic
Ferro-Tic	Brass	Beryllium Copper
CPM 10V	Cold Roll Steel	Hastalloy
4130	Hot Rolled Steel	Stellite
Graphite	Stainless Steels	Titanium

Machining costs are usually determined by the amount of square inches of cutting (as illustrated in Figure 2:39), type of material, programming, set up, and whether the flushing nozzles are on the part.

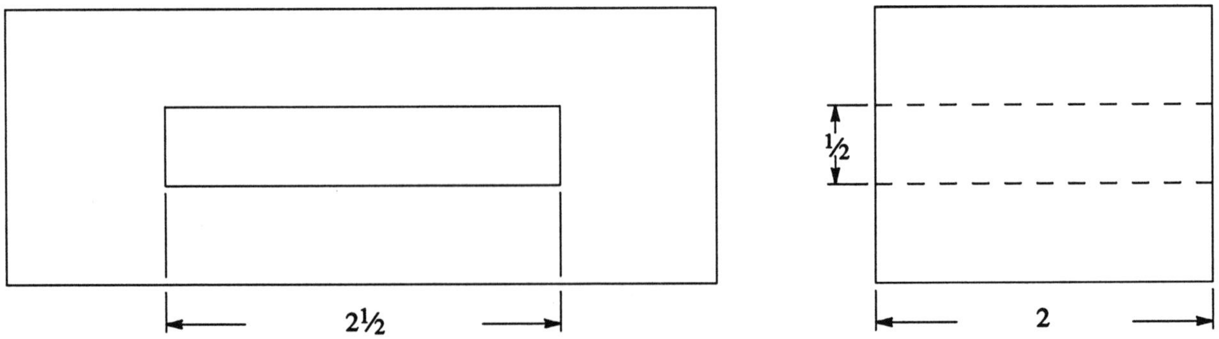

Figure 2:39 Determining Machining Costs

Thickness x Linear Inches = Square Inches
2 x (1/2 + 1/2 + 2 1/2 + 2 1/2) = 12 Square Inches

Entering the New World of Wire EDM Machining

Wire EDM is here to stay due to its cost effectiveness. Designers should be aware that their use of wire EDM will enable them to improve their product by making changes which would have either been impossible with conventional machining or cost prohibitive. Today they can fine tune their tolerances and redesign their parts without adding significant costs to their product.

3

Proper Procedures

3 Proper Procedures

To gain the greatest benefits from wire EDM, specific procedures should be established to maximize EDM's potential in reducing machining costs.

In planning your work, visualize the wire EDM machine as a super precision band saw which can cut any hard or soft electrical conductive material.

Starting Methods for Edges and Holes

Three methods to pick up dimensions.

The part to be EDMed should be indicated in if the outside edges are important.

1. Pick Up Two Edges as in Figure 3:1.

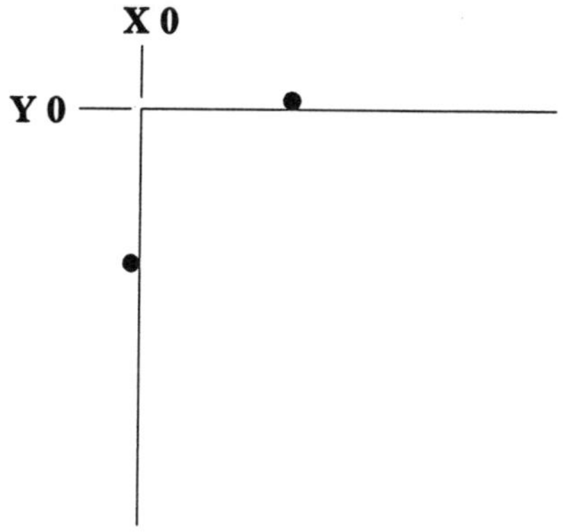

Figure 3:1

2. Pick Up Hole as in Figure 3:2.

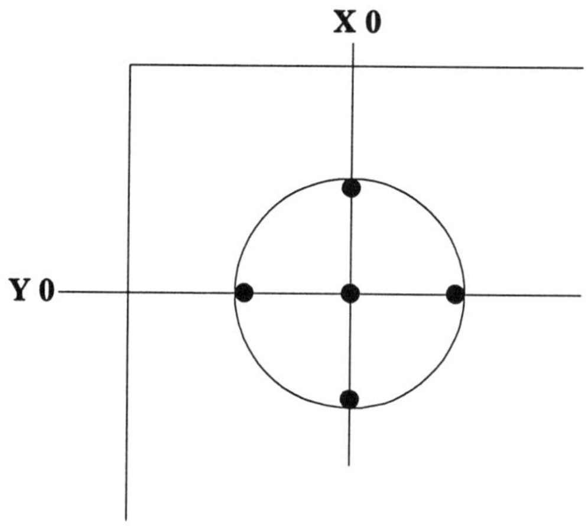

Figure 3:2

3. Pick Up Edge and Holes, or Two Holes as in Figure 3:3.

By using an edge and two holes, a part which is much larger than the capacity of the machine can be EDMed. The part is indicated in and a hole that has been either machined or EDMed is picked up. Additionally, two EDMed edges can be used to pick up after the part is moved.

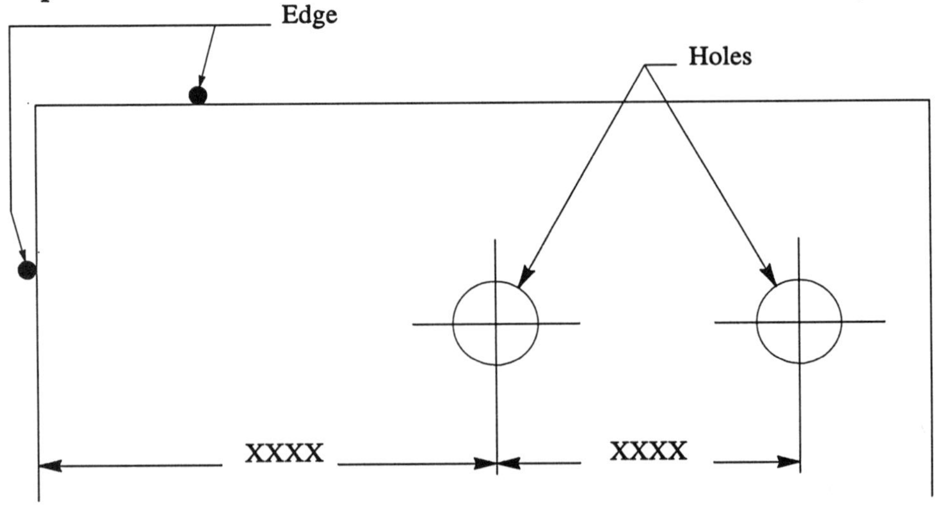

Figure 3:3 Pick-Up From Edges and Holes

Edge Preparation

A. Square Edges

1. Machined or Ground. To ensure accuracy, the pick up edges must be square as shown in Figure 3:4.

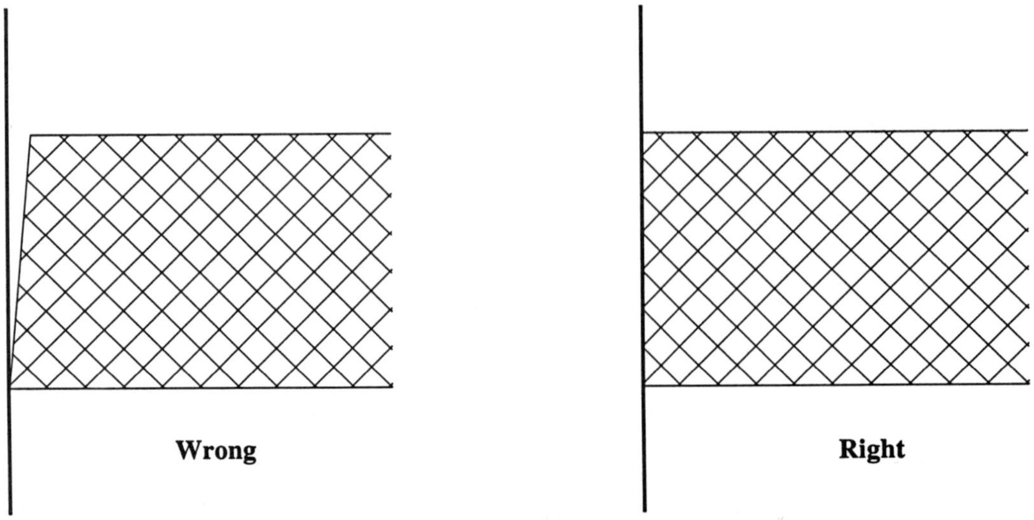

Figure 3:4 Edges must be square for proper pick up

2. Unfinished Edges. Parts can be made square to top surface with sides unfinished as illustrated in Figure 3:5. The wire is made square to the top surface by means of a special squaring block.

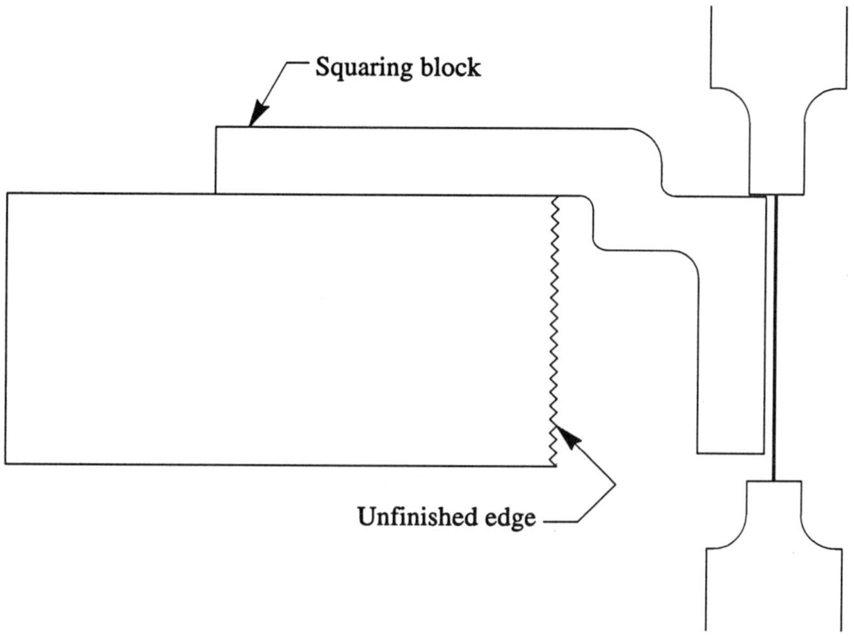

Figure 3:5 Squaring block can be used to make wire square to surface of material to be cut

B. Scale

Since wire EDM is an electrical process, any material that is non-conductive must be removed. Scale from heat treating, as in Figure 3:6, is non-conductive. Make sure that heat treated parts, particularly holes, are either cleaned of scale or have been vacuum heat treated or wrapped.

Sand or glass blasting can be used to clean the surfaces where the wire will cut in. However, it is difficult to clean deep or blind holes with sand or glass blasting.

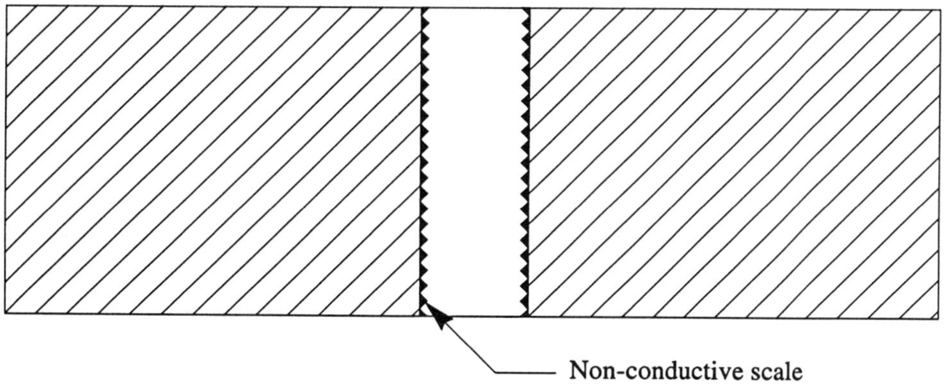

Figure 3:6 Hole must be free of scale to pick up holes

C. Pick-up

To properly pick up a surface, there should be no obstructions as in Figure 3:7. If necessary, pick-ups can be made from a step by means of a gage block or gage pin. See Figure 3:8.

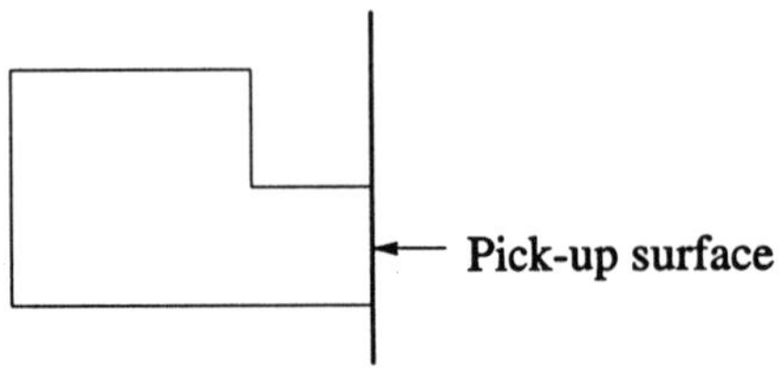

Figure 3:7 Non-obstructive Pick-up

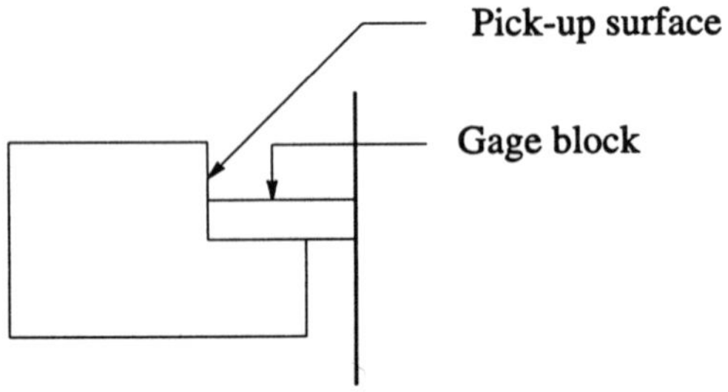

Figure 3:8 Gage Block Used for Pick-up

Starter Holes

A. Automatic Pick-up

When a starter hole is used for locating the part, the machine will automatically pick up the center of the hole as shown in Figure 3:9. Make sure scale from heat treating has been removed.

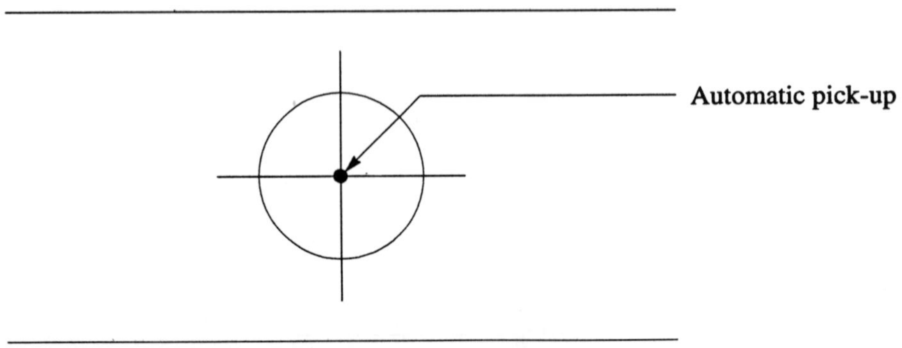

Figure 3:9 Wire EDM machine will automatically pick up the center of a hole

B. Unsquare Holes

If the hole is not square, as illustrated in Figure 3:10, the wire will pick up the high points and not the center of the hole.

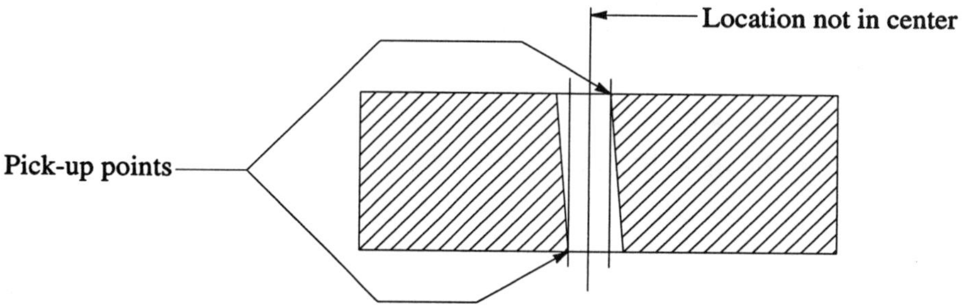

Figure 3:10 Unsquare hole will produce inaccurate pick-up

C. Relieved Holes

A relieved hole, as pictured in Figure 3:11, is the most accurate method to pick up from a hole. Leave approximately 1/8" to 1/4" of land.

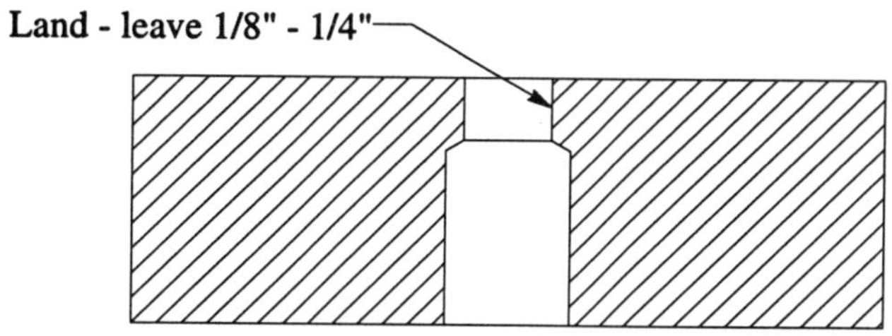

Figure 3:11 The greatest accuracy is obtained with a relieved hole

D. Smooth Holes

A drilled hole may leave ragged edges. The wire will pick up the high points of the ragged edge. To ensure accuracy, a reamed or bored hole is best. See Figure 3:12.

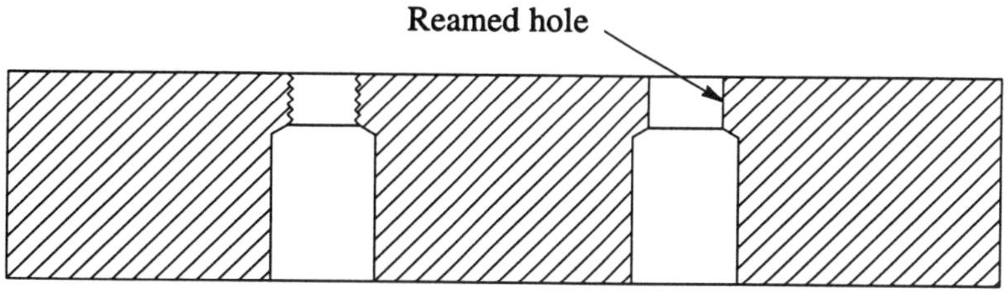

Figure 3:12 Smooth reamed holes are the most accurate for pick-up location

E. Placement and Location of Starter Holes

1. If the pick-up is at another location, the starter hole does not need to be accurately located.

2. Place the starter hole at a straight surface whenever possible as shown in Figure 3:13. When parts are not skim cut, usually a slightly raised area appears where the part ends. In that case, a file or a stone can be used to remove the tip.

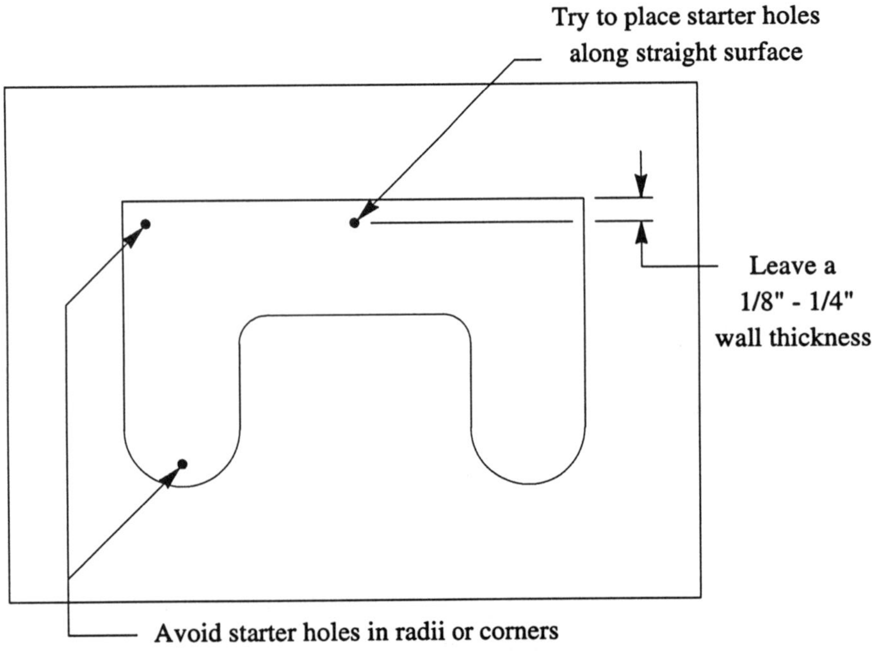

Figure 3:13 Proper Placement of Starter Holes

3. On thin cuts, place the starter hole in a corner, as illustrated in Figure 3:14, so that only one slug will be made.

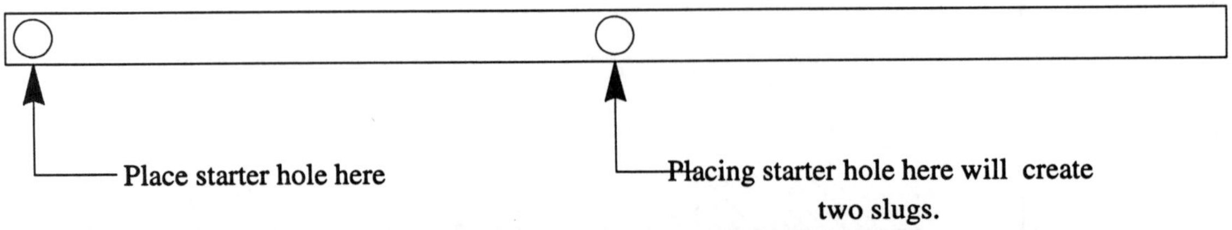

Figure 3:14 Proper Placement of Starter Hole for Thin Cuts

Layout

If multiple wire EDM operations are made in one piece, the best method to put in dimensions is from a reference point of X = 0, Y = 0. See Figure 3:15 for the ideal layout.

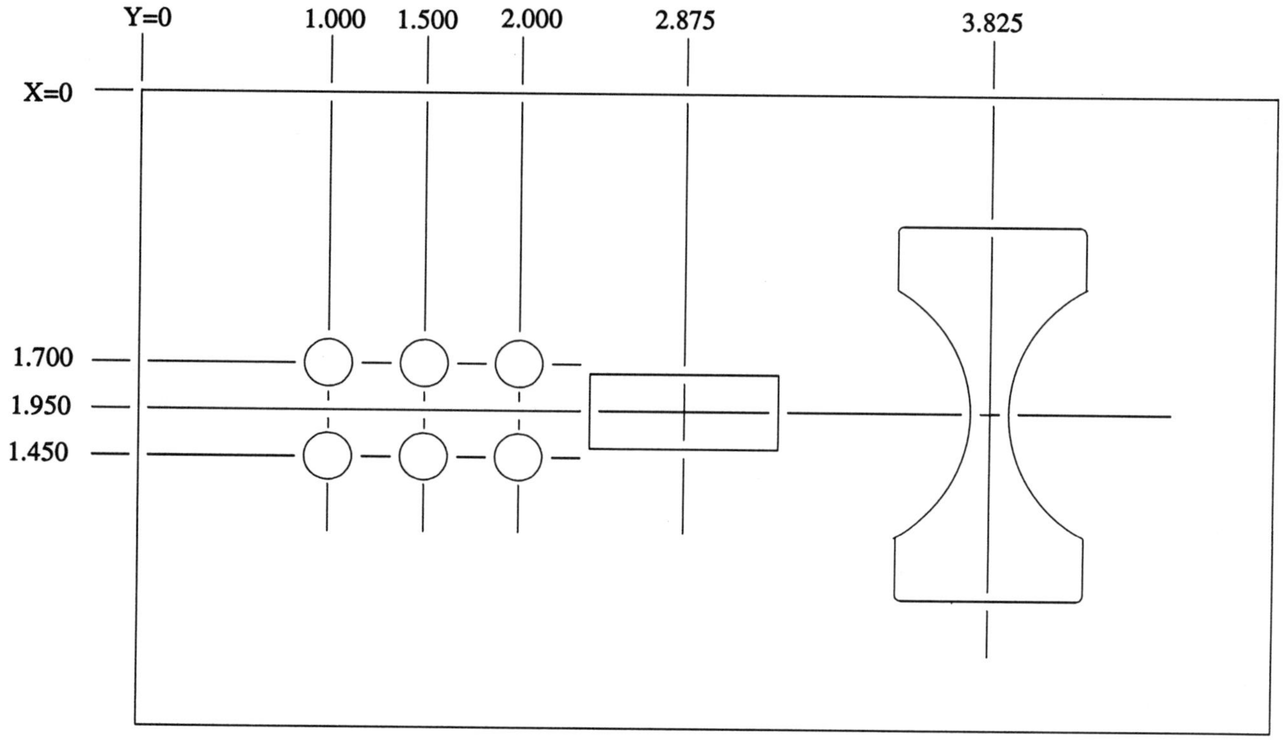

Figure 3:15 Best Layout Dimensions for Wire EDM

4

Understanding the Wire EDM Process

4 Understanding the Wire EDM Process

Accuracy and Tolerances

Wire EDM is extremely accurate. Machines move in increments of 40 millionths of an inch (.00004), some in 10 millionths of an inch (.00001), and some even in 4 millionths of an inch (.000004). Machines can hold +/-.0002; however, on thin pieces they can hold +/-.0001. To obtain such tolerances, skim cuts have to be made. See Figure 4:1.

Courtesy Mitsubishi Electric Corporation

Figure 4:1 Precision Wire EDMing

Finishes

An extremely fine finish can be produced with wire EDM, in some machines below 15 RMS. Even in the so called "rough cut," wire EDM produces an excellent finish. When customers are shown the cuts made from a one pass cut, they are often amazed at the fine finish.

This fine finish is present even when very large parts are cut as in Figure 4:2. In other cutting operations, such as lasers and water jet, the larger the part, the rougher the finish. The reason for the smooth finish with wire EDM is that the wire electrode goes through the entire part and spark erosion takes place along the entire wire electrode.

Figure 4:2 Part 16 Inches Long Cut at Reliable EDM Corporation

Wire Path

A. Wire Kerf

The wire never comes into contact with the workpiece. The wire electrode cuts by means of spark erosion and leaves a slightly larger path than the wire. A commonly used wire is .012, which usually creates a .015 to .016 kerf as shown in Figure 4:3. Thinner wires have smaller kerfs.

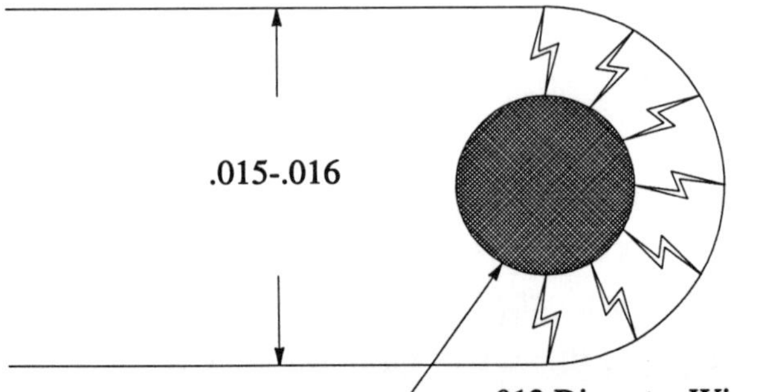

Figure 4:3 Wire Kerf

B. Corners and Radii

When the wire turns a corner, it can leave a sharp edge on the outside corner, but it will always leave a small radius on the inside corner as demonstrated in Figure 4:4. The size of this radius is determined by the wire diameter plus spark gap. It is cost effective to avoid sharp corners whenever possible. To produce very sharp corners, skim cuts must be made. In dies, sharp corners become the area most subject to wear, and in other applications a small radius is often preferable.

The minimum radius for .012 wire is .007, for .010 wire, .006. Smaller radii are possible with thinner wire; however, because thinner wire cuts slower, most work is done with the larger wires.

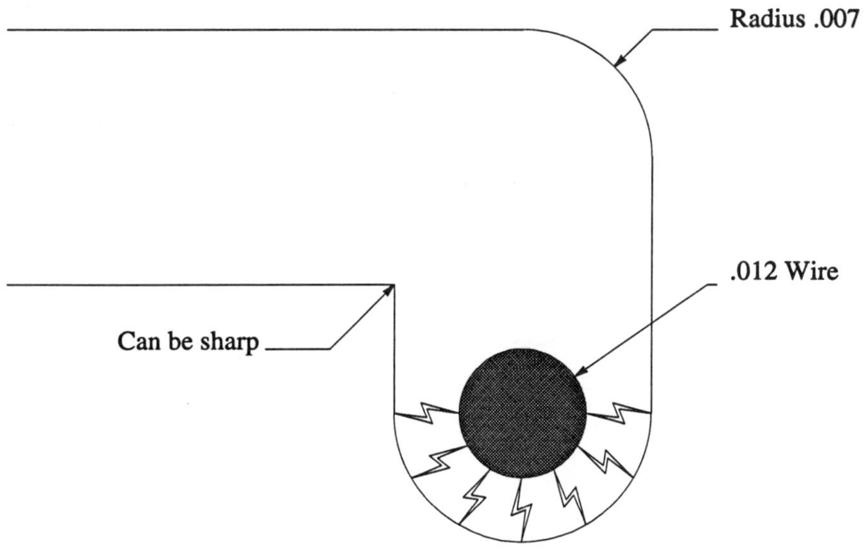

Figure 4:4 Inside and Outside Corners

Skim Cutting

For most jobs the initial cut is more than sufficient for both finish and accuracy. For precision parts, sometimes skim cuts are performed to get greater accuracy and a finer finish. There are four main factors in the need for skim cuts.

A. Barreling Effect and Wire Trail Off

There is a .001 to .002 gap between the wire and the workpiece. In this gap a controlled localized eruption takes place. The force of the spark and the gases trying to escape causes a slight barreling effect. On thick workpieces it causes the center to be slightly hollow. See Figure 4:5.

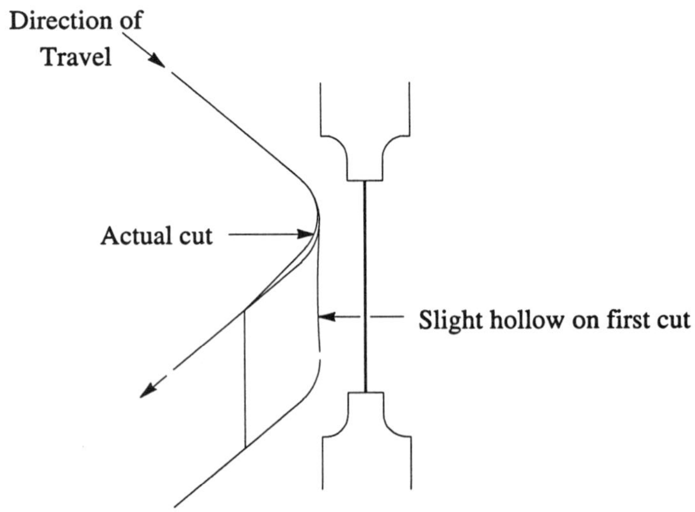

Figure 4:5 First Cut Corner Conditions

When cutting sharp corners, the wire dwells longer by the inside radius, causing a slight overcut; on the outside radius, it speeds, leaving a slight undercut as illustrated in Figure 4:6.

A trail off is produced when the machine cuts a corner. A slight amount of material is left behind for a short distance before the wire returns to its programmed path. For most jobs this slight undercut is negligible.

The sharper the corner the greater the overcut and undercut. The accuracy of the part determines the need for skim cutting.

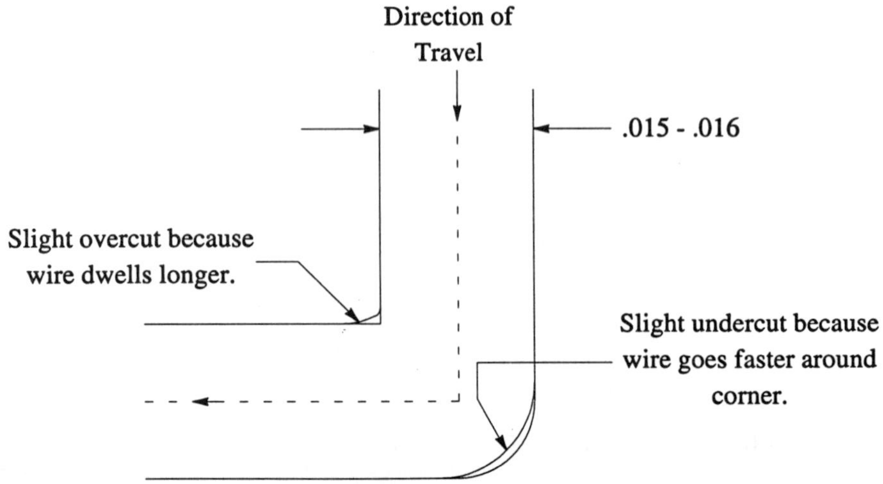

Figure 4:6 Skim Cutting Is Used For Very Close Tolerances

B. Metal Movement

Even though metal has been stress relieved, the metal sometimes moves when the part is cut because of the stresses that were initially in the material. On precision parts this requires skim cutting to bring the part into tolerance. The number of skim cuts are determined by the accuracies called forth by the print.

C. Finishes and Accuracy

First cuts produce a fine finish; however, sometimes a finer finish and greater accuracy are required. To accomplish this, skim cuts are used. Customers are often amazed at the fine finish from a high-speed first cut. See Figure 4:7 for various finishes that can be produced with wire EDM.

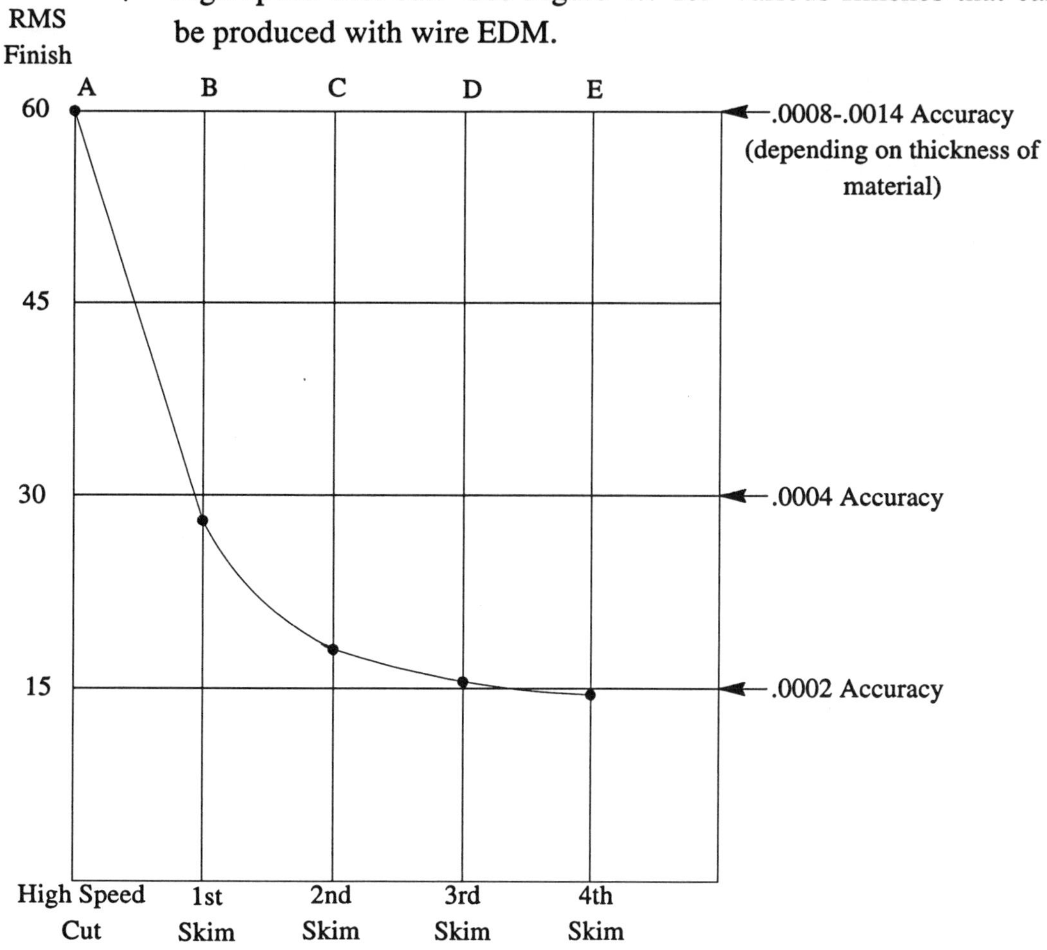

Figure 4:7

Cut A - For most jobs this finish and accuracy are more than adequate.

Cuts B-E - Sepending on accuracy and finish required, various numbers of skim cuts are performed.

D. Carbide

Carbide is a non-conductor. It is the binder or matrix that is eroded away. As the binder is EDMed, the carbide nodules fall out. The energy applied during the cutting determines the depth of binder that is removed.

On some carbide jobs, one cut is sufficient. However, on many occasions skim cuts are used to receive a better surface quality. Today, some machines are equipped with AC power supplies. These machines are especially beneficial when cutting carbide in that they produce smaller heat-affected zones and less cobalt depletion than DC power supplies.

Flushing

Flushing is an important factor in wire EDM. Flushing pressure is produced from both the top and bottom flushing nozzles as illustrated in Figure 4:8. The pressurized deionized fluid helps in the production of the spark and in the removal of the eroded metal particles.

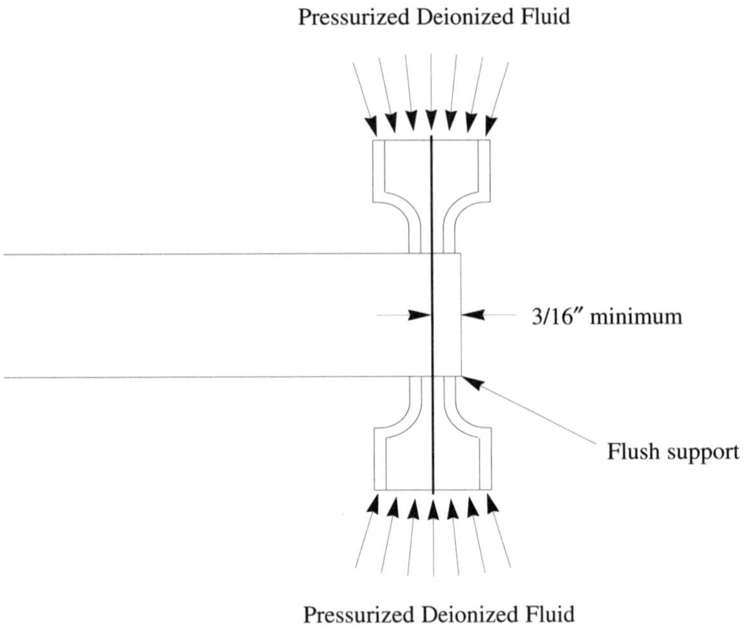

Figure 4:8 Ideal Flushing Conditions

Sometimes customers bring in work in which the flushing nozzle comes close to the edge of a workpiece as shown in Figure 4:9. When this occurs the flushing pressure is lost and sometimes the wire breaks and accuracy is effected. To avoid wire breakage, a lower spark energy is used, thus slowing the machining process.

Whenever possible, at least 3/16" of material should be left to support the flushing nozzles.

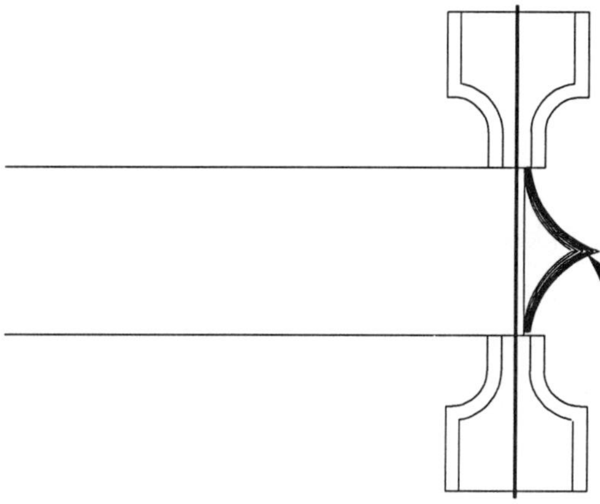

Figure 4:9 Poor Flushing Conditions

Cutting Speed

Speed is determined by the square inches of cut per hour. Manufacturers rate their equipment under ideal conditions: usually 2 1/4 inch thick hardened D2 tool steel with perfect flushing conditions. Differences in piece thickness, varying materials, and producing sharp corners can greatly alter the speeds of these machines.

The conductivity and the melting properties of materials determine their cutting speed. Aluminum is a good conductor and has a low melting temperature--it cuts much faster than steel.

Carbide, on the other hand, is a non-conductor. The binder, which is often cobalt, is melted away which causes the carbides to fall out. Carbide cuts much more slowly than steel. Various carbides cut at different speeds. The size of the carbide grains and the amount and type of binder determine the cutting speed.

Impurities

Generally there is little difficulty with impurities; however, occasionally materials are received with non-conductive impurities. The wire electrode will either stall or pass around the impurity if it is small. This can cause streaks which may leave a raised or indented surface.

Welds can be cut, but all slag must be removed where the wire will cut. Tig welding is preferred for wire EDM.

Recast

Recast was a problem in the past on the older capacitor type machines. However, today's newer machines, depending on the power settings and skim cuts, leave from .0003″ on the first cut to no recast. An independent metallurgical laboratory inspected a titanium aerospace part made on a Mitsubishi, and under 1000x magnification, they found no recast layer.

On some machines today, the heat effected zone is generally .0015″ on the first cut, .0003″ on the first skim cut, and .0001″ on the second skim cut. Since with ram EDM much more energy can be applied to the electrode, ram EDM produces a much greater heat effected zone. At Reliable EDM Corporation many jobs are produced with one cut.

Heat-Treated Steels

Wire EDM will cut hard or soft material. In fact, material in the hardened condition will cut slightly faster. Materials requiring heat-treatment are most commonly treated before being cut with wire. This eliminates the distortions generally created by heat-treating.

Cutting Large Sections

Steels from the mills have inherent stresses. Even hardened steel that has been tempered has stresses. For small cutting sections, the effect is negligible; however, when large sections are to be cut, and close tolerances are needed, it is advisable to remove some of the metal.

Cut out the major portion of metal as in Figure 4:10, or put in relieving slots to connecting holes with a band saw to remove the stresses as in Figure 4:11. Steel should be stress-relieved before heat-treating to remove the stresses caused by the milling, drilling, and grinding. After heat-treating, the tool steel should be double or triple drawn, including the non-deforming air hardening tool steels. Another method to remove stresses is to use cryogenics (deep freeze). The tool steel is hardened and tempered, then it is put into deep freeze and retempered. It is best that the top and bottom surfaces of the plate are ground.

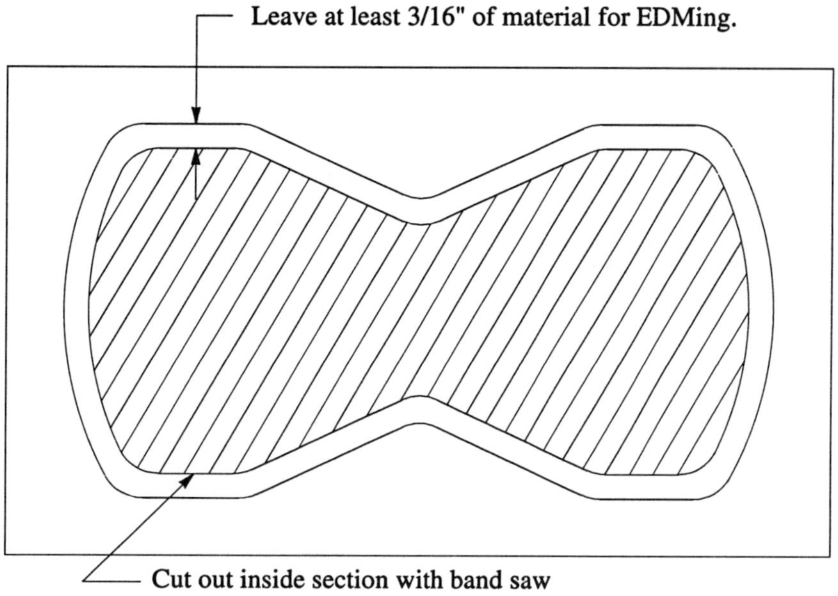

Figure 4:10 Removing Material to Reduce Stresses on Large Parts

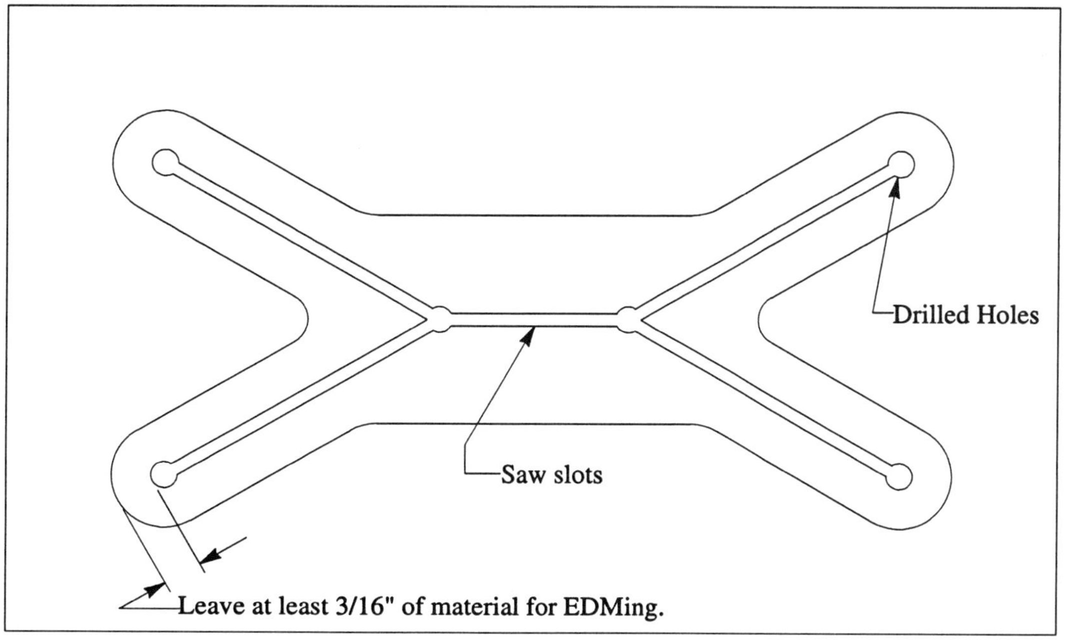

Figure 4:11 Put in saw slots to reduce stresses on large parts

Cutting Sections From a Block

A. Leaving a Frame

When a section must be cut from a block of steel, leave a frame around the workpiece to ensure accuracy and to reduce cost. Leave at least 1/4" to 1/2" around the part so the flush nozzles can properly remove the eroded particles and support the part when clamping as illustrated in Figure 4:12.

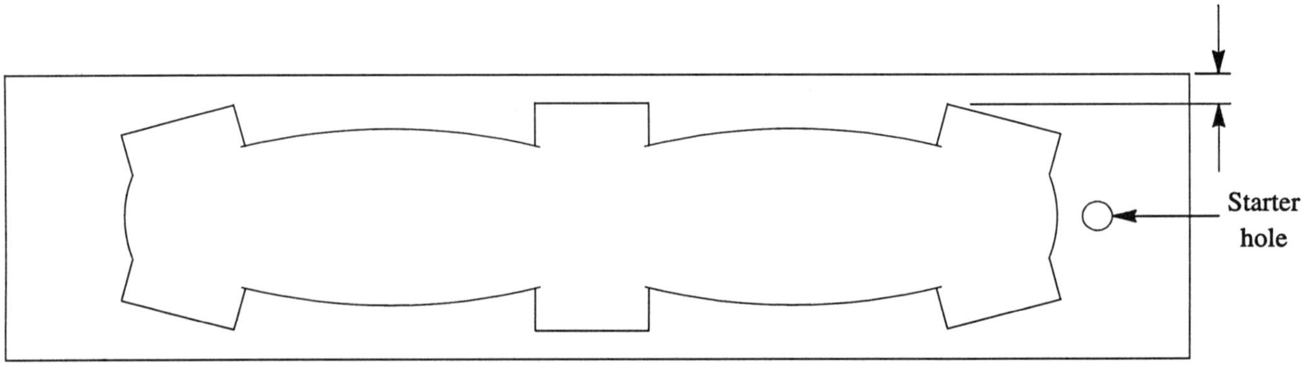

Figure 4:12 Support Part with Frame

B. Strength of Frame

Leave sufficient material around the part so that when the part is held in a fixture, it does not move after it is EDMed. Figure 4:13 demonstrates a weak support frame.

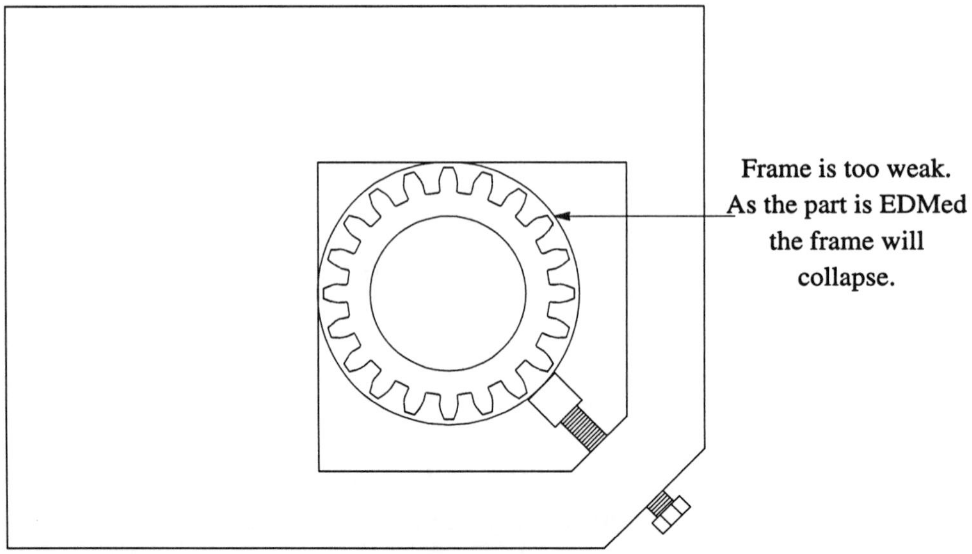

Figure 4:13 Improper Support Frame

C. Material for Clamping

For some parts, provision should be made for clamping. All parts must be clamped in some manner. See Figure 4:14.

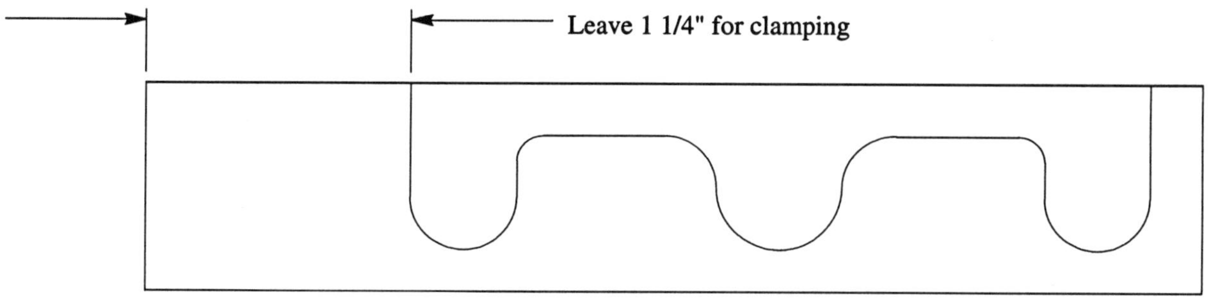

Figure 4:14 Certain parts need extra material for clamping

Understanding the Wire EDM Process

Most problems are not with the wire EDM, but rather with the characteristics of the particular material being cut. The better understanding one gains of the wire EDM process, the more beneficial this amazing process becomes. Let us look at some of the ways to reduce costs.

5

Reducing Costs

5 Reducing Costs

Wire EDM costs can sometimes be greatly reduced if the material has been prepared properly.

Create One Slug

The creation of one slug should be targeted. Wire EDM is an automatic process; if more slugs are made, more down time and operator services are required.

When cutting surfaces which are close to an edge, inadequate flushing occurs which reduces cutting speed. When entering a workpiece, lower power settings and slower cutting speeds must be used due to the poor flushing. When entering a part which causes a feathered edge, some slight surface irregularities will appear. Skim cutting can be used to remove the irregularities; however, doing unnecessary skim cuts increases cost. Cutting one slug is much more cost effective. See Figs. 5:1-4.

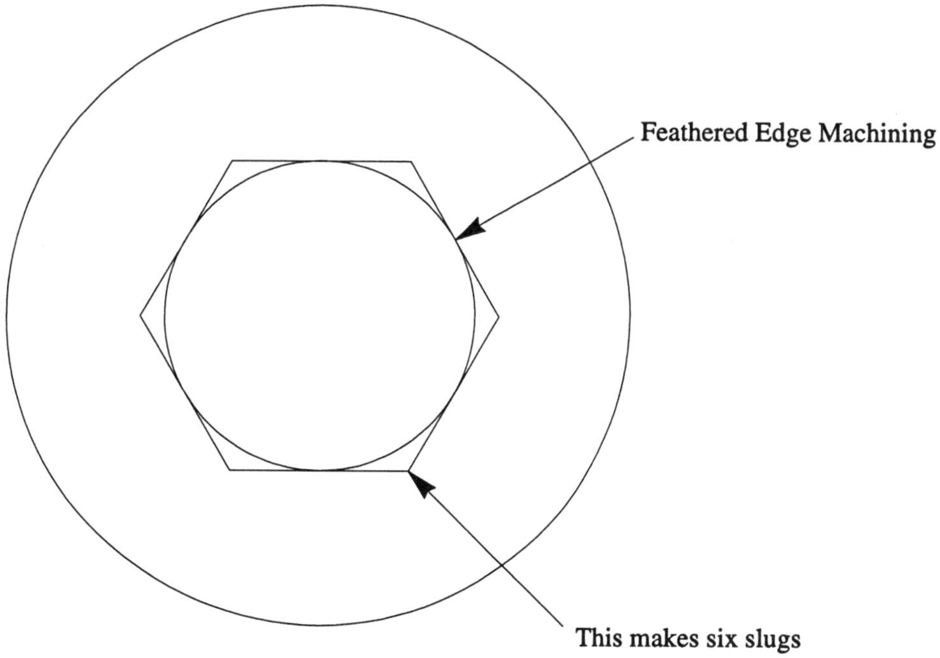

Figure 5:1 Wrong
Having hole on size for a hex cut creates six slugs plus slower machining due to feather edge machining.

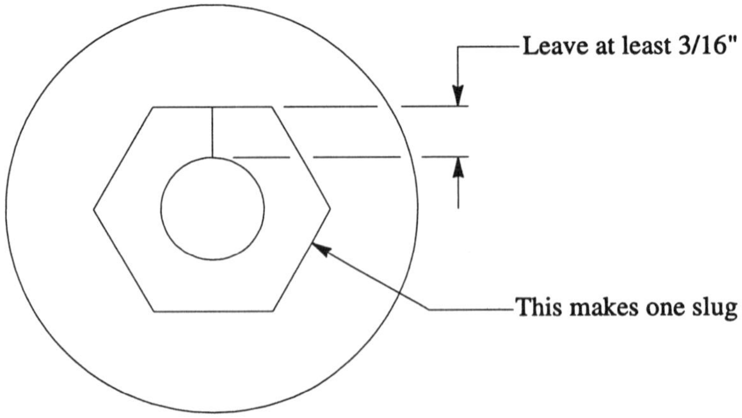

Figure 5:2 Right
Putting in a small hole creates one slug and allows for much more efficient machining.

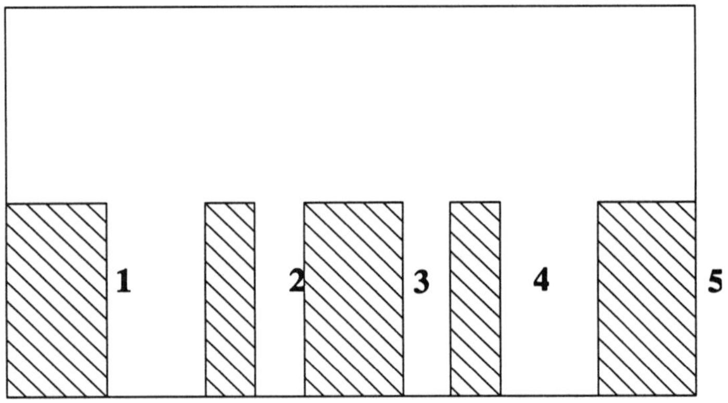

Figure 5:3 Wrong: Creates Five Slugs
Five starting cuts must be made, and five times the machine must be stopped to remove the slugs.

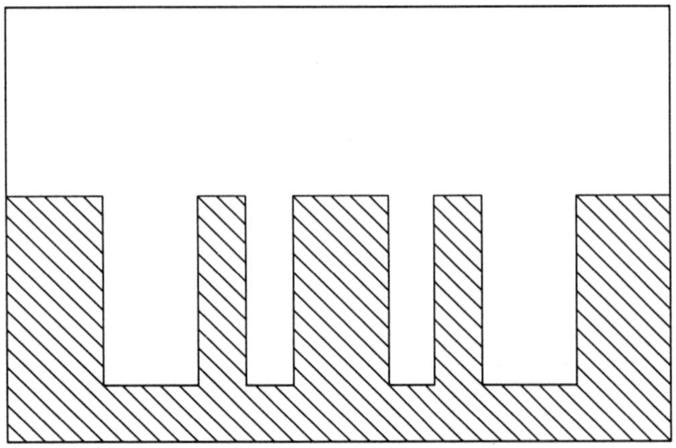

Figure 5:4 Right: Creates One Slug
Leaving material on the outside allows one slug to be made.

Cutting in the Air

There is no problem cutting in the air with wire EDM as illustrated in Figure 5:5; however, the most efficient method is when both top and bottom flush ports are on the workpiece as shown in Figure 5:6.

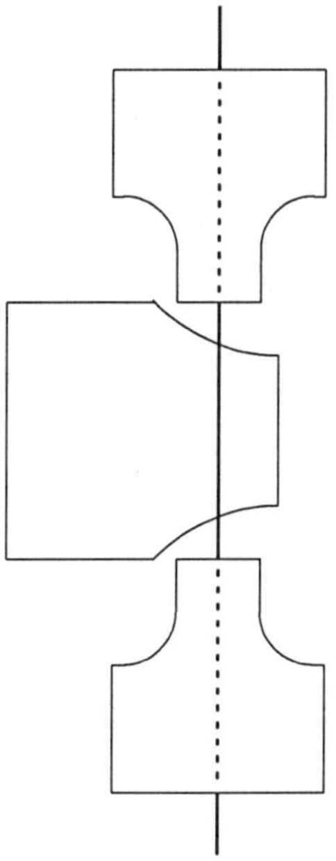

Figure 5:5 Air Machining: Can be done but less efficient

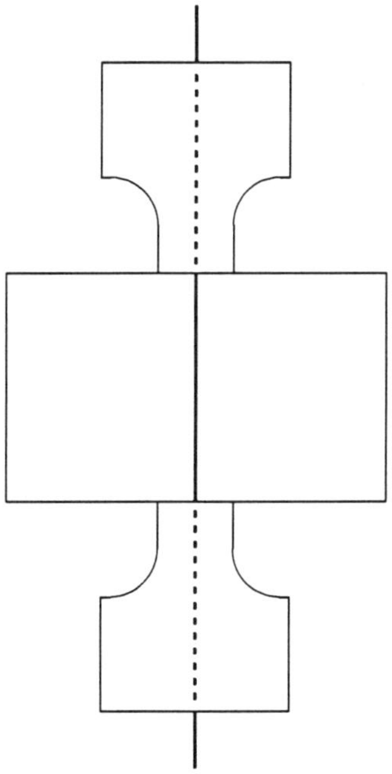

Figure 5:6 Most efficient cutting condition when both flush ports rest on part

Machining After Wire EDM

Sometimes it is much more cost effective to do the machining after the EDM process rather than before, as demonstrated in Figure 5:7. Often parts are stacked to reduce costs. When parts have intricate dimensions, stacking may be difficult if parts have been previously machined as shown in Figures 5:8,9.

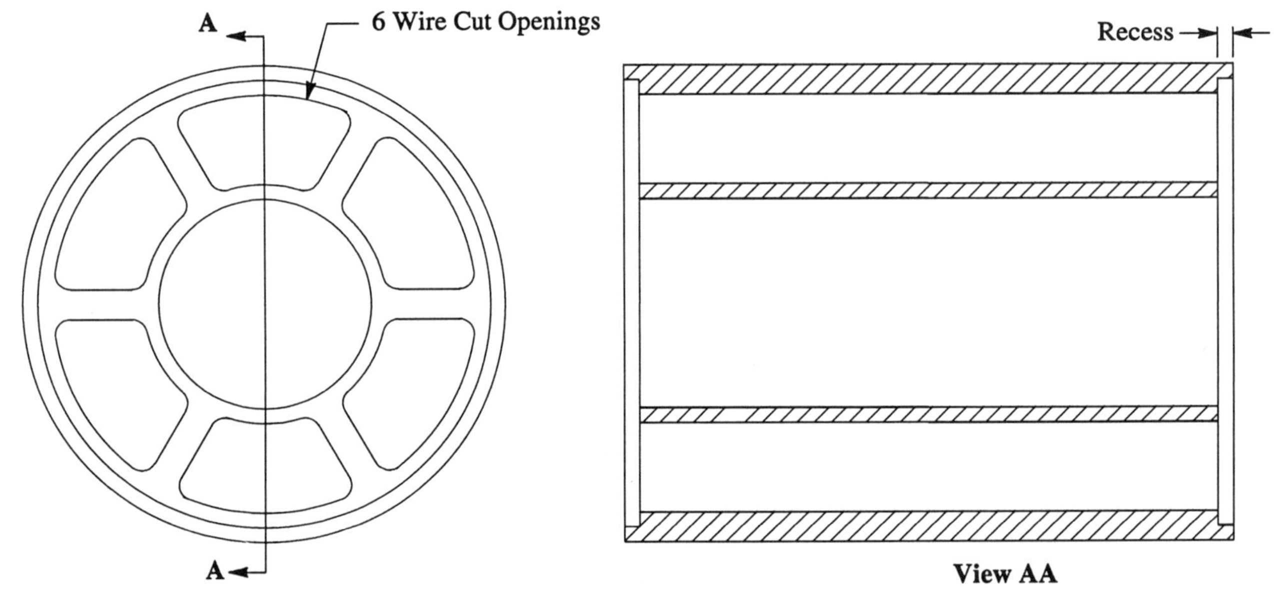

Figure 5:7 Machine after EDMing
Since recess is shallow, it is more cost effective to do EDMing when block is solid.

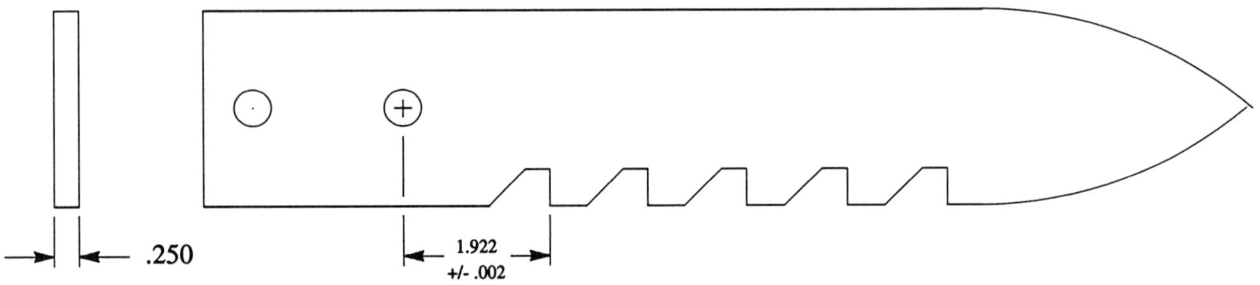

Figure 5:8 Put holes in after EDMing
Making one piece presents no problem; however, parts like these are stacked. If holes are pre-machined, it is difficult to line up both holes when cutting large stacks.

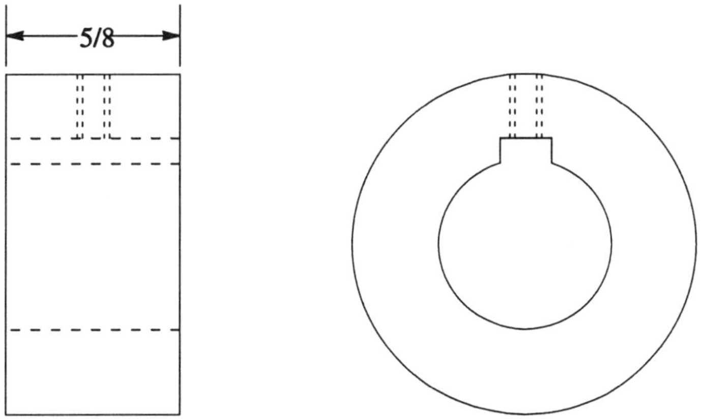

Figure 5:9 Put tapped hole in after EDMing

These parts get stacked in a "V" block. Tapped holes cause alignment difficulties and higher costs because lower power settings are needed due to air pockets in holes.

Multiple Plate and Sheet Metal Parts

Stacked sheet metal can be held with fixtures without welding. However, when multiple parts are needed from one stack, and starter holes are needed, it is often best to bolt with flat head screws or weld the stack. Make sure stack is flat and inquire for the ideal thickness for stacked parts. Accuracy, efficiency, and machine capabilities determine height of stacked parts.

Wire EDM will cut through light rust, but heavy rust and scale must be removed. Many times plates are warped. Clamp the parts tightly together before welding. Leave at least 1/2" on the sides for welding and clamping the part. See Figure 5:10 for proper stacking.

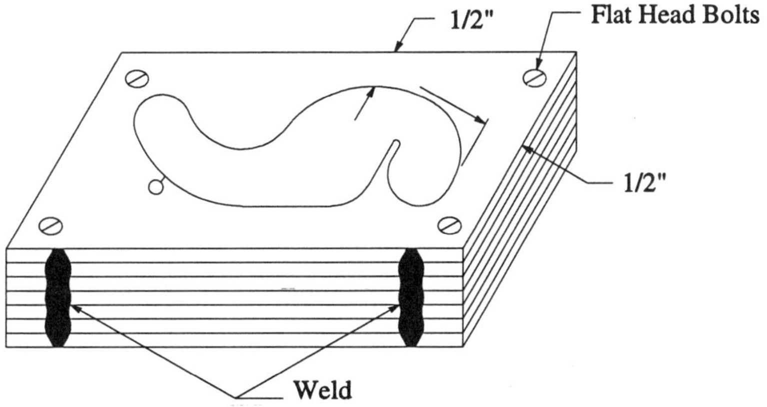

Figure 5:10 Welded Stacks

Leave at least 1/2" for clamping and a frame to support part while cutting. Caution: If parts are bolted or welded, make sure both sides of plates are clean and free from heavy scale, tape, paper or any other non-conductive materials.

If sheets or plates are badly warped, take half of each stack and let the belly hit the center. Clamp the ends together and weld. Try to produce a straight surface. Remove weld from top and bottom of stack so flush ports do not hit the weld.

In welding, be particularly careful that all the sheets are clean--remove all marker paint, (not magic marker), scale, tape or paper between the sheets. Wire EDM cuts by spark erosion; it cannot cut through non-conductive materials.

Production Lots

Wire EDM is an excellent machining method for production work. Fixtures are often used to hold the parts. It is important that production lots are machined the same in the area where they will be located. Parts also need to be machined square. See Figure 5:11.

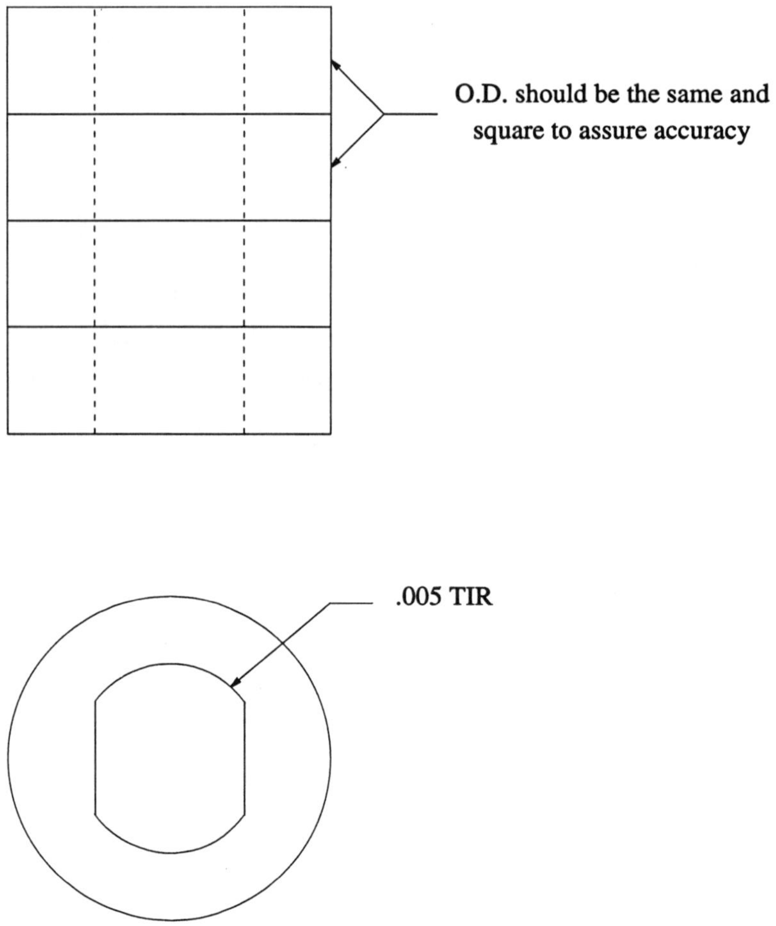

Figure 5:11 Production EDMing
Consideration should be made for stacking parts.

Pre-machining Non-complicated Shapes

It is not always necessary to EDM the entire part. Sometimes pre-machining can reduce costs as shown in Figure 5:12.

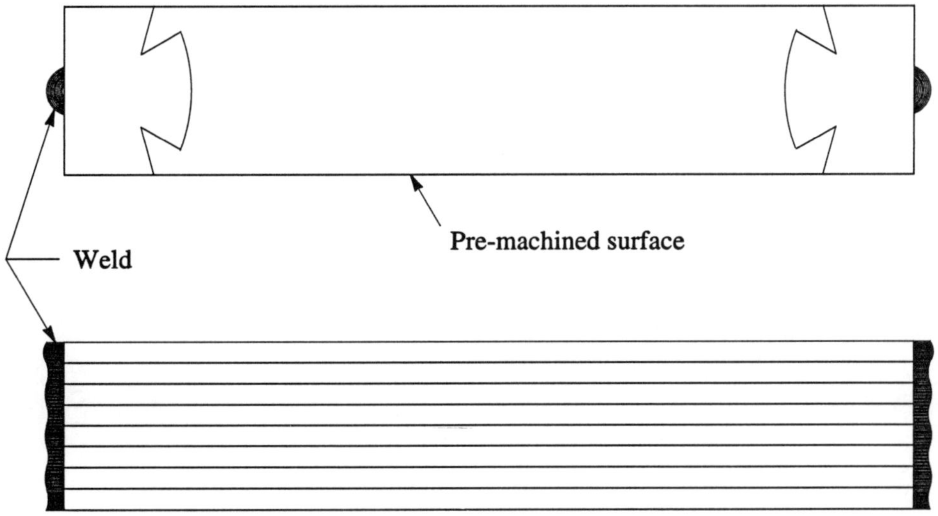

Figure 5:12 Pre-machine Parts to Reduce Costs

Wire EDM is an extremely efficient method to machine parts. There are, however, many ways in which the costs of this process can be further reduced.

The introduction of wire EDM into the manufacturing process has greatly altered the machining process for many parts, and the greatest impact has been in the tool and die shop.

6

Production Wire EDM

6 Production Wire EDM

Wire EDM can no longer be considered non-traditional machining. In the last ten years, cutting speed with wire EDM has improved dramatically. Now, in many applications, cost effectiveness dictates that parts be cut with wire EDM rather than by conventional methods.

When Reliable EDM Corporation was started, the foremost advantage of using wire EDM was first thought to be in extrusion molds and dies. However, it was soon discovered that Reliable's largest customers were production machine shops. As machine shops discovered the benefits of wire, they increasingly sent many of their parts to be wire EDMed. Today, the bulk of Reliable's business is production EDM.

Defining Production EDM

Visualizing production, some immediately think of thousands of parts, but any multiple number of parts can also be considered production. See Figures 6:1-5.

Figure 6:1 Precision Cams Cut From Multiple Stainless Sheets

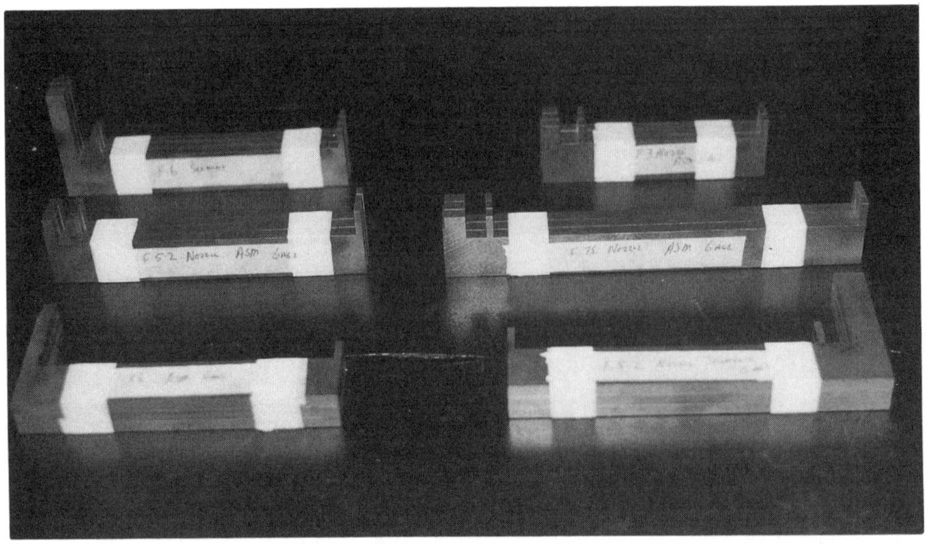

Figure 6:2 Short Run Production of Multiple Gages

Figure 6:3 Production Cutting of Machine Parts From 17-4 Stainless Steel

Figure 6:4 Special Fixture to Cut Multiple Machine Parts

Figure 6:5 Titanium Parts for Oil Exploration

The Greatest Problem

The greatest problem Reliable EDM Corporation has encountered is that many designers and engineers are not aware of wire EDM potential. As more and more engineers, tool designers and machinists discover the many benefits of wire EDM, they increasingly turn to job shop wire EDM companies. One of the major tasks Reliable EDM has undertaken is to educate the industry in the capabilities and advantages of wire EDM.

Outsourcing or Owning a Wire EDM

Those consistently using wire EDM undoubtedly have considered purchasing a wire EDM machine. This may be prudent for some, however for many others, there are many benefits of using job shop EDM companies.

Advantages of Job Shop Wire EDM Companies

A. Competitive Outsourcing

Quotes to various companies can ensure the lowest cost per part.

B. Quality

If parts are not to print specifications, the job shop is held responsible. In outsourcing, quality is guaranteed.

C. Service

Shops with multiple wire EDM machines have the capability for quick turn around.

D. Efficiency

Job shops must remain competitive in order to stay in business, and consequently keep abreast of latest technologies. Companies, with one or two of their own wire EDM's often have one operator, who often does not have the expertise to maintain maximum efficiency of the wire EDM machines.

E. Fixtures

Job shops can do in minutes what would otherwise be an elaborate, time-consuming setup in-house. At Reliable EDM Corporation numerous fixtures and holding devices are at hand to reduce set up costs.

F. Inspection Equipment

Some shops are not equipped with precision inspection equipment and MIL-STD-45662 and MIL-1-45208A inspection requirements. Comparators and new high-magnification video measuring microscope systems are now available to measure parts accurately as illustrated in Figure 6:6.

Figure 6:6 Reliable EDM's Video Measuring Microscope System

This system is able to magnify parts to 325X.

G. Machine and Operating Costs

Wire EDM machines are costly. Following are some costs that should be considered:

Five Year Estimated Costs	Cost
Cost of Wire EDM Machine	
Loan Interest	
Shipping	
Rigging	
Property Taxes	
Fixtures	
Inspection Equipment	
Additional Room	
Air Conditioning	
Attending Classes at Suppliers Training Center	
Weeks of In-House Training	
Months of Possible Errors From Inexperience	
Maintenance of Machines	
Wire	
Filters, power Contacts, and resin	
Breakdown of Machines	
Out of Town Repair Services	
Electricity	
Machine Not Used Due to Economic Slowdown	
Programmer/Operator Costs	
Other Expenses	
Total Cost	
Value of Machine After Five Years	
Actual Cost	
Divide by 5	
Estimated Yearly Cost	

Cost and profit-conscious businessmen should carefully calculate the costs of operating and maintaining wire EDM machines before making a decision to purchase. Reliable EDM Corporation's experience has been that when operations are competitive, few companies purchase their own machines.

Reliable EDM Corporation is founded on the principle of providing customers with quality work, dependable service, and competitive pricing; and in nine years, it has become the largest wire EDM job shop west of the Mississippi.

H. Competitive Pricing

Some tool and die shops have in-house wire EDM machines due to a constant need for making their punches and dies. Often they are not as competitive as a production wire EDM shop which specializes in doing wire EDM. Some shops have older machines, and though they cut accurately, these machines are slow in comparison to newer models.

I. Altered Machining Methods

Wire EDM has indeed changed many machining methods. As designers, engineers, machinists, and others in the metal machining field become more familiar with Electrical Discharge Machining, many more applications for this process will be found.

7

Various Wire EDM Applications

7 Various Wire EDM Applications

Examples of EDM Uses

Once the capabilities of Wire EDM are understood, many unique applications can be made. Reliable EDM is the largest wire EDM job shop west of the Mississippi. And as such, its machinists have EDMed thousands of jobs for numerous customers.

As designers, engineers, and machinists become increasingly familiar with EDM capabilities, they will see even greater utilization of this process. Following are some examples showing various applications.

A. Tall Parts

Four angular cutouts were made on a large shaft (See Figure 7:1).

Figure 7:1 Flange Cutout
Slug is on top of flange.

At our company we have modified a number of our machines to cut taller parts. Illustrated in Figure 7:2 is a machine we modified to cut 19 1/2" (495 mm). To the right of it is a 30" (762 mm) bar of titanium we wire EDMed on another machine. Due to customer confidentiality, we are not showing a 37" (940 mm) cut we did in a bar of beryllium copper. We modified this machine to cut 38" (965 mm).

Figure 7:2
Machines Modified: Left--Cut 19-1/2 Inches Tall (495 mm)
Right: Titanium Cut 30 inches tall (762 mm). Ruler: 36" (914mm)

Submersible Cutting of Tall Parts

We also did a slight modification on another machine to cut 18" (457 mm) submersed. See Figure 7:3.

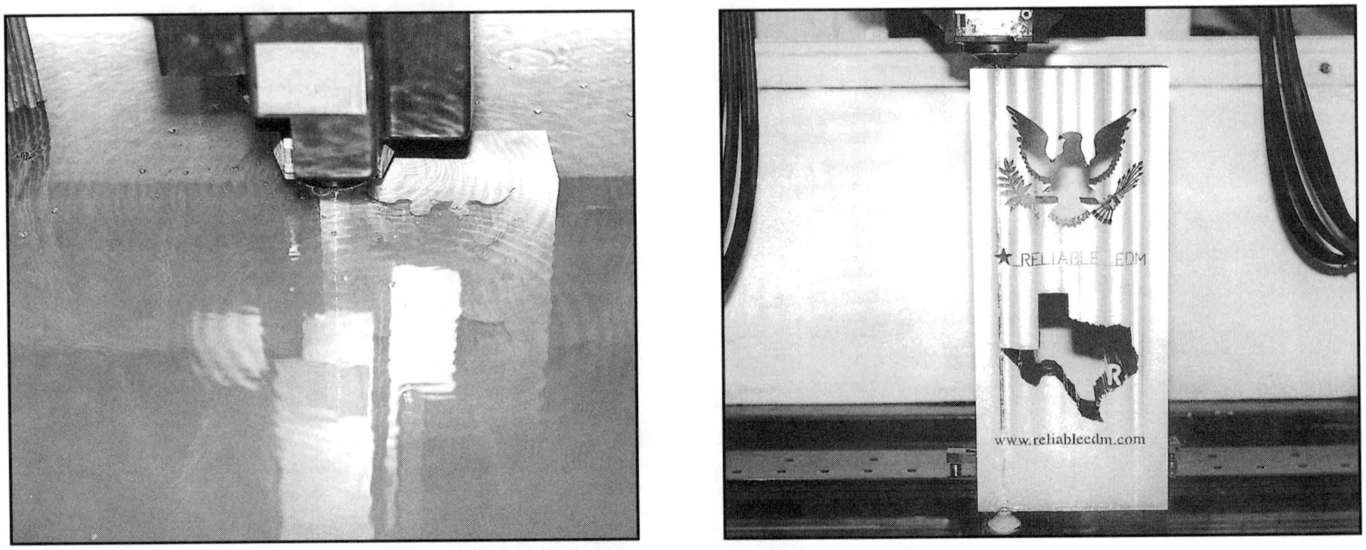

Figure 7:3
Submersible Cutting to 18" (457 mm)

A large 24 inch mold was cut. The part was then moved and recut. See Figure 7:4.

Figure 7:4 A Large Mold
The ruler is 24 inches.

B. Overhanging Parts

Figure 7:5 Cutting A Special Shear Blade

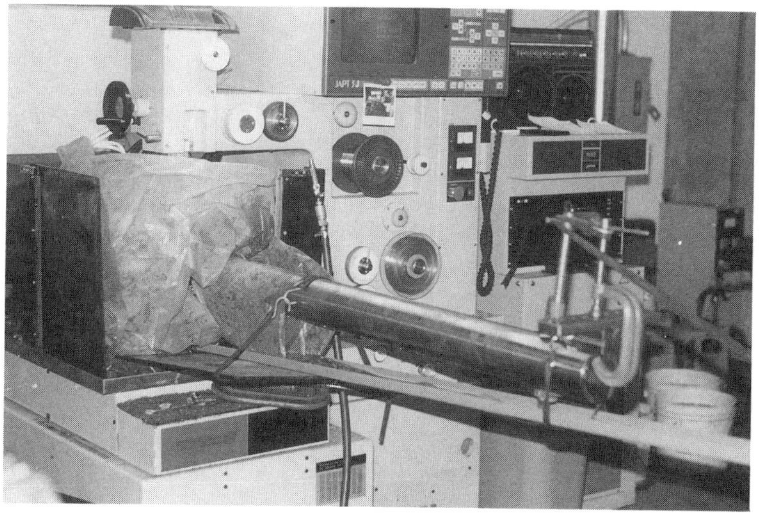

Figure 7:6 Splitting A Long Tube

C. Other Applications

Figure 7:7 Repairing A Mold

Figure 7:8 Cutting The Outside Diameter Of A Gate Valve

Figure 7:9 Cuts Made From The Side Of A Tube

Figure 7:10 One-cut: Inside Diameter And Keyway

Figure 7:11 Candle Mold Parts Split In Half

Figure 7:12 Test Specimen From A Large Gear

Figure 7:13 Finished Test Specimen Cut
Flush Nozzles Were Never In Contact With Gear (11 1/2 Inch Thick Cut).

Wave of the Future

Wire EDM is here to stay. As designers, engineers and machine shop supervisors discover the many possibilities of this revolutionary machining method, more and more parts will be produced in this fashion.

Unit Two

The Tool and Die Revolution

8

Advantages of Wire EDM for Die Making

8 Advantages of Wire EDM for Die Making

Old-Fashioned Tool and Die Making

Wire EDM has revolutionized tool and die making. In 1950, Carl Sommer started to work in a machine shop; one year later he became an apprentice tool and die maker in a large handbag frame plant in Brooklyn. There, they produced a large variety of handbag frames, which required many different kinds of fixtures and dies.

From 01 tool steel they milled, ground or filed the form punch. The punch was then hardened in a gas-fired oven without a pyrometer. In those days one learned early the necessary cherry red color indicating that the punch was ready to be quenched in oil. Then with a torch they often tempered the punch to a light straw color.

Using the hardened punch they traced it on a piece of tool steel. A band saw was used to carefully cut away from the line. The hardened punch was placed on the top of the soft die section and placed under a power press. By hand the press was bounced until an indentation was made. A filing machine and hand files were used to file the excess material away. Dykem blue was brushed in the cavity, and the punch and die were placed in the press again. Then back to the filing machine. This process was repeated over and over until a proper fit was made. The filing machine was set on an angle to produce the proper taper. The die was then hardened with hopes that the 01 tool steel would not distort when quenched in oil.

After working there a number of years, Sommer began working in a precision die shop in Long Island City, Queens N Y. The precision die shop in Long Island City was a new world of die making--here the die sections were precisely ground. He ground some die sections within .0001. Days were often spent on the surface grinder, constantly on guard lest a part would be scrapped.

In making these dies, no comparator or optical equipment was used. The first worker used a large magnifier to check for the clearance; this made his eyes bloodshot red. When Sommer's turn came to make these dies, he decided to grind the punch and die sections to precise dimensions. Instead of constantly relying on sight, he used a tenth indicator to obtain the proper dimensions. This is one of the notes he wrote in Figure 8:1: "Grind flat with .016 radius. Move cross feed .001 at a time. Leave .0003 over for finish grinding. Then touch front and back off .0002, then grind flat. Use dresser three times."

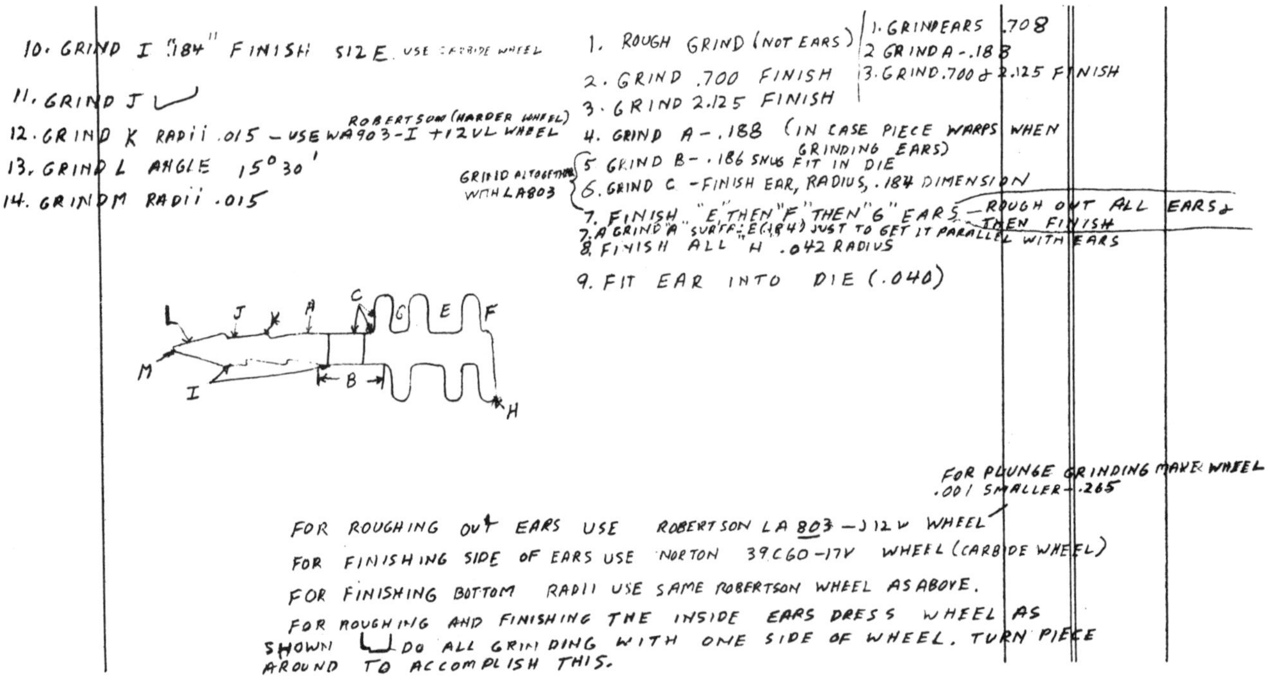

Figure 8:1 Author's shop handwritten sketch for grinding floral pick punches and dies

These dies ran continuously. The picks went into automatic dispensers, and no burrs were tolerated. The called-for clearance was between .0005 to .001.

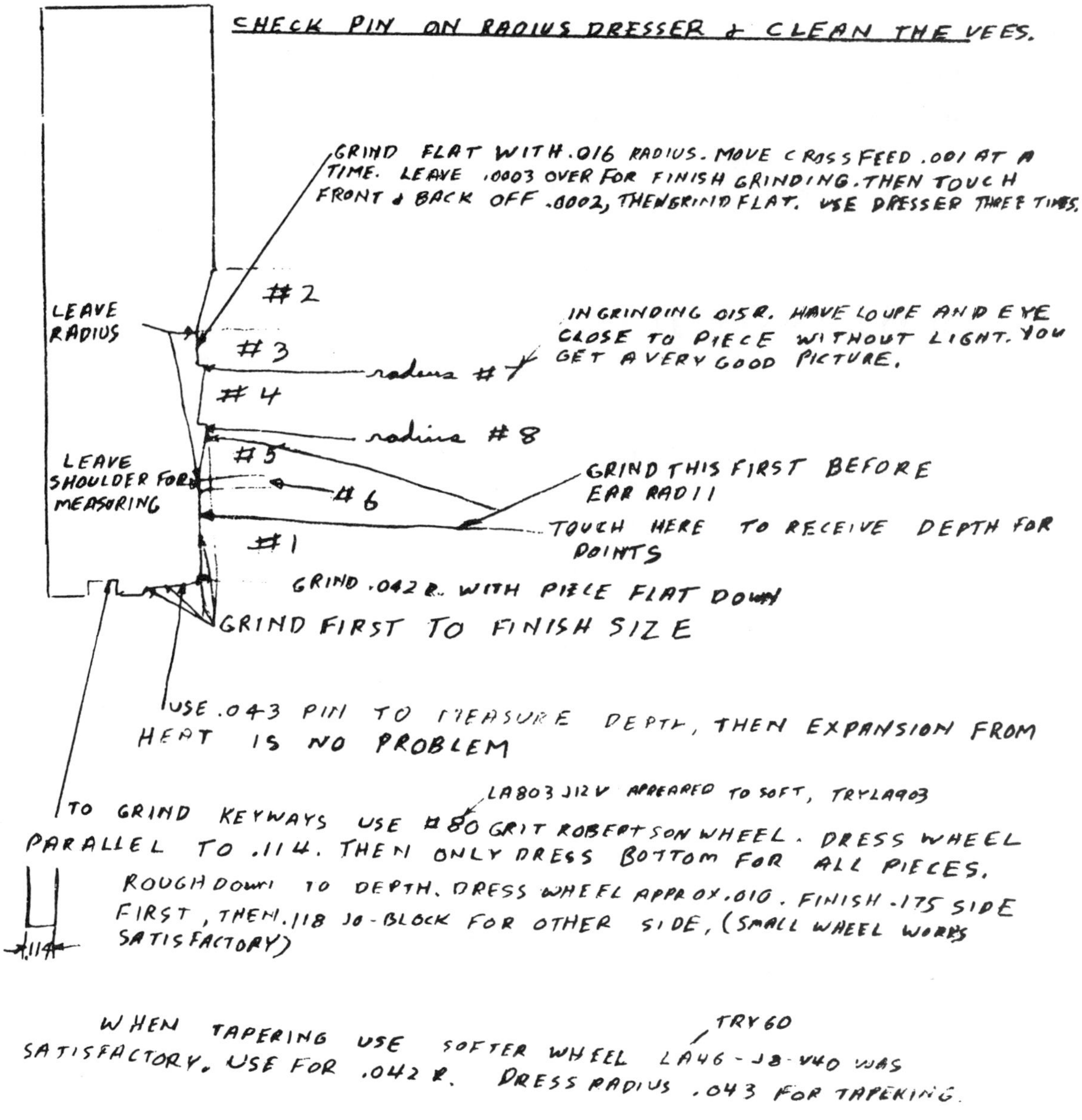

Figure 8:2 Old-Fashioned Precision Die Making

One note: Grind flat with .016 radius. Move crossfeed .001 at a time. Leave .0003 over for finish grinding. Then touch front & back off .0002, then grind flat. Use dresser three times.

The Revolution

Highly skilled tool and die makers were necessary to produce these kinds of tools. Then came wire EDM. By making a computer program of the shape, production of a better and a more accurate tool was possible.

Tool and die makers are still needed to assemble tooling, but wire EDM has eliminated the need for those skilled die makers to make all the elaborate punch and die sections. Today, that costly and laborious job is done by wire EDM. As a result, tooling costs have been greatly reduced, and at the same time a better quality die has been produced.

Advantages of Wire EDM Dies

A. One-Piece Die Sections

Previously complicated dies were sectionalized--this allowed the die sections to breathe. Now with wire EDM, the die can be made from a solid block which results in a much more rigid die. Sectionalized dies require much more mounting time than a one-piece die section.

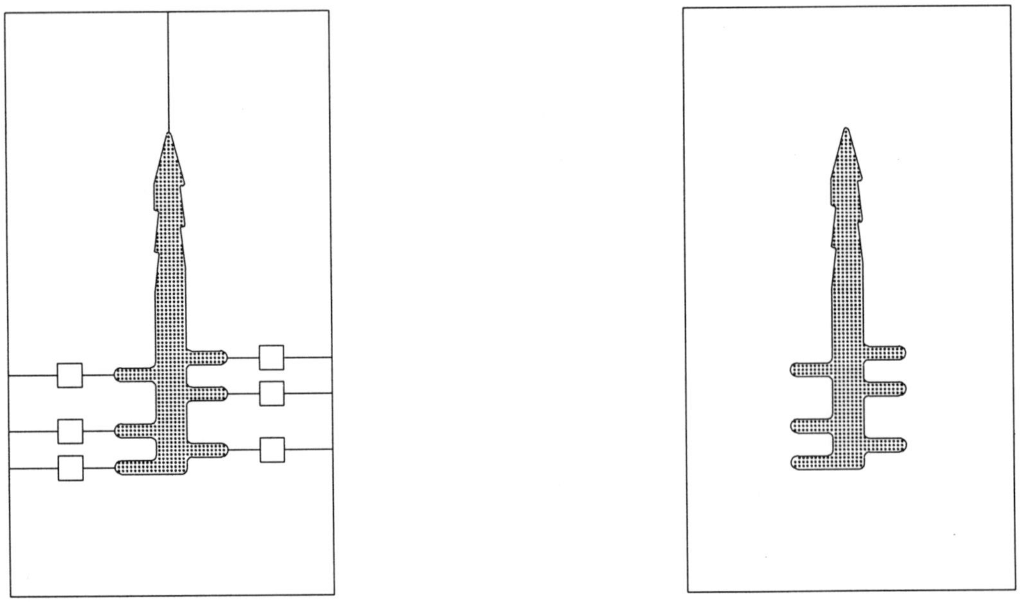

Figure 8:3 Sectionalized Die Sections **Figure 8:4 Solid Die Section**

Wire EDM has eliminated the costly sectionalized die for the superior and less costly solid dies.

B. Spare Parts

To keep up production, spare sections can be on hand in case of wear or breakage. Spare sections can be precisely duplicated without having the previous part. Previous programs can be stored in the computer, on tape or floppy disc.

C. Dowel Holes EDMed

When die sections or punches are likely to be changed due to wear or design change, the dowel holes can be EDMed. All that is needed is a starter hole in the approximate location, and all parts can be precisely machined to the dowel holes.

D. Better Tool Steels

Dies were often made from 01, an oil-hardening tool steel, because it was easier to machine and grind. With wire EDM, the better performing air-hardening tool steels are used, resulting in much longer tool life.

E. Accuracy

Many wire EDM machines move in increments of at least 40 millionths of an inch (.00004); therefore accurate forms and clearances can be easily maintained.

F. Die Repairs

Broken dies can be saved by replacing the damaged section with a wire EDMed insert, or the damaged area can be hard welded and wire EDMed. See Figures 8:5,6.

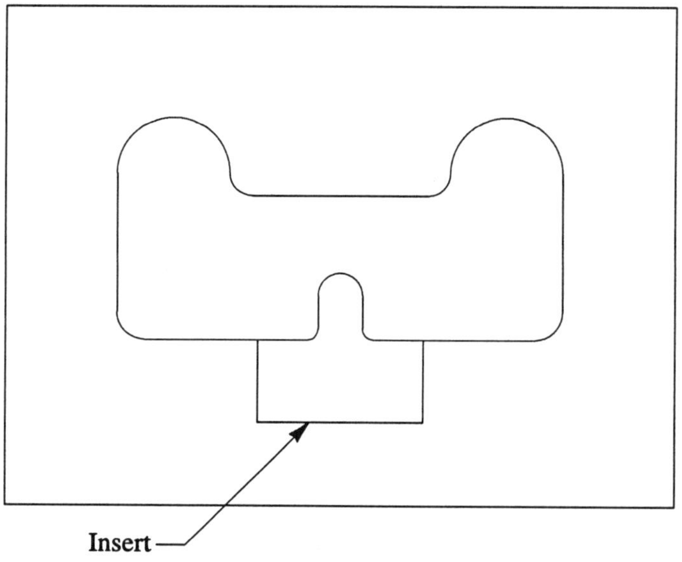

Figure 8:5 Damaged Die Section Repaired with an Insert

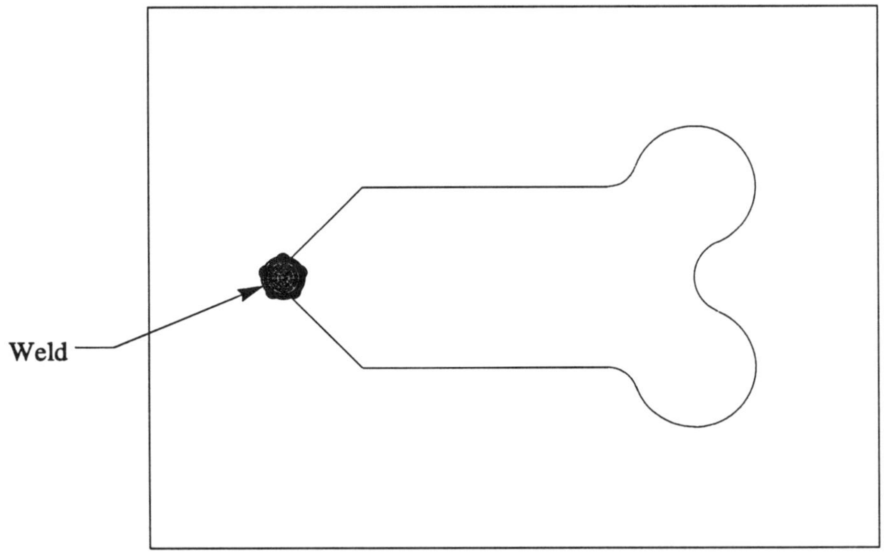

Figure 8:6 Damaged point in die section can be hand welded and then wire EDMed to its former shape.

G. Fine Textured Finish

It has been discovered that the fine textured surface produced from wire EDM produces longer tool life because of its ability to retain lubricant.

H. Eliminates Distortion

Punch and dies can be wire EDMed after heat-treatment. This eliminates the distortions created in heat-treating.

I. Inserts for High Wear Areas

If certain areas have a much larger wear ratio in the die, inserts can be designed. Instead of sharpening the entire die, inserts can be installed.

J. Smaller Dies

Costs for dies made with wire EDM may be reduced because smaller progressive dies are possible.

K. Longer Lasting

A die lasts only as long as its weakest link. Because die clearance is precisely maintained and wire EDMed dies can be ground down much farther due to the precision tapers, they often last longer than conventional tooling.

L. Punches and Dies From One Piece of Tool Steel

Punch and die sections can be produced from one piece of tool steel. (Chapter 10 provides complete detail regarding the production of these long lasting inexpensive dies.)

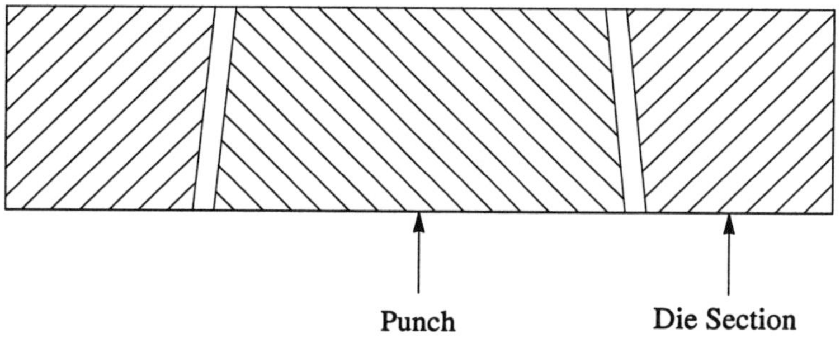

Figure 8:7 Punch and Die Made from One Piece of Tool Steel

M. Cutting Stripper and Die Section Together

The stripper can often be mounted on the bottom of the die section and cut with the die section. This significantly reduces the cost of making the stripper alone.

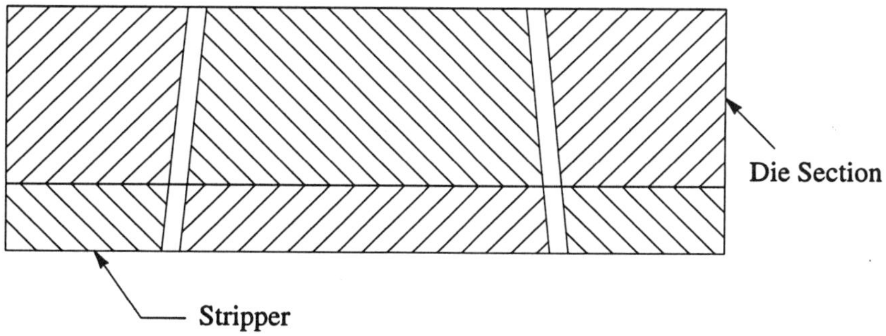

Figure 8:8 Cutting Stripper and Die Section Together

Wire EDM has truly revolutionized the tool and die making industry. Let us examine the many ways to take advantages of this new technology.

9

Wire EDMing Punch and Die Sections

9 Wire EDMing Punch and Die Sections

Punches

A. Large Punches

Large punches, as shown in Figure 9:1, can be held by putting dowels and screws into the body of the punch. Use large enough dowels and particularly large enough screws to prevent the punch from moving in case of a misfeed.

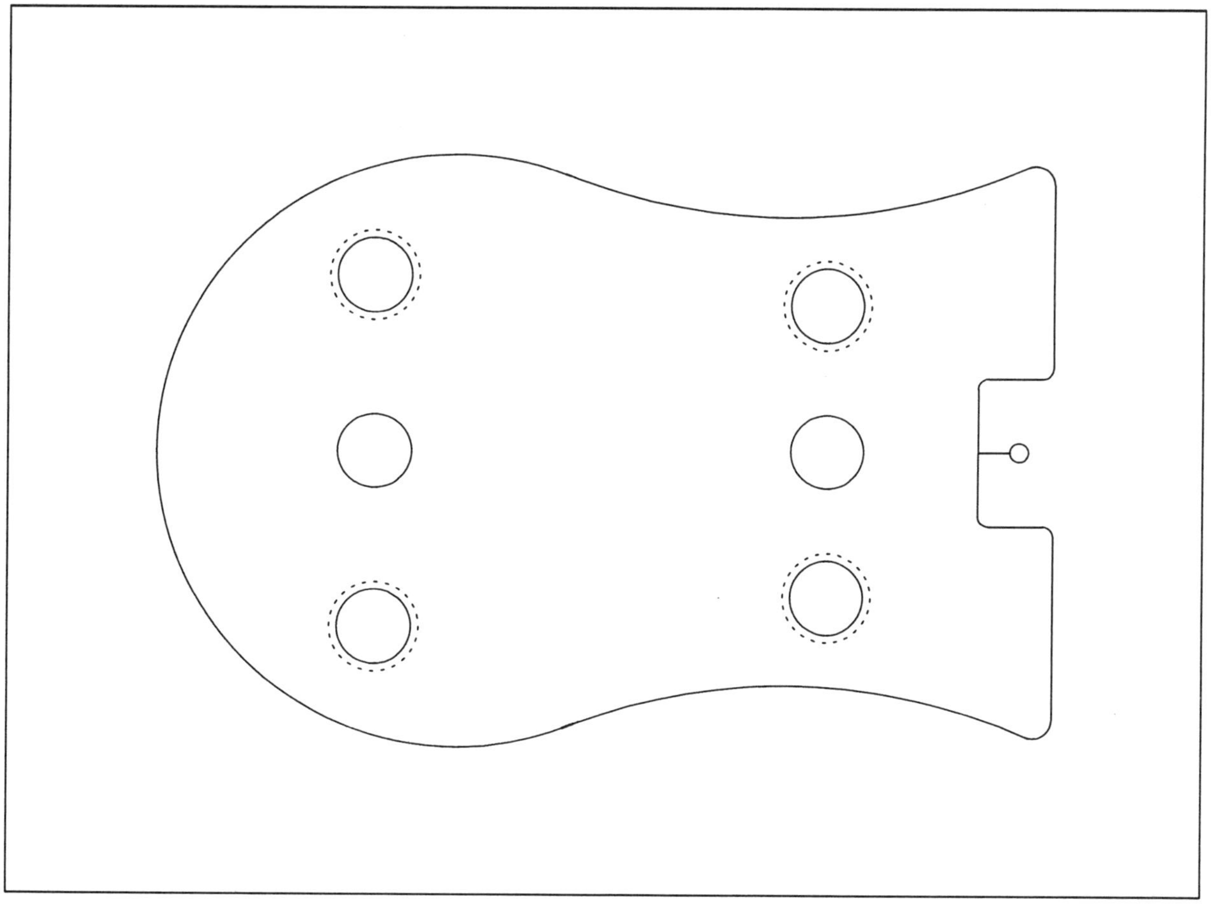

Figure 9:1 Large Punches
Bolt large punches directly to punch holder.

B. Small Punches

There are various methods of holding small punches as illustrated in Figure 9:2.

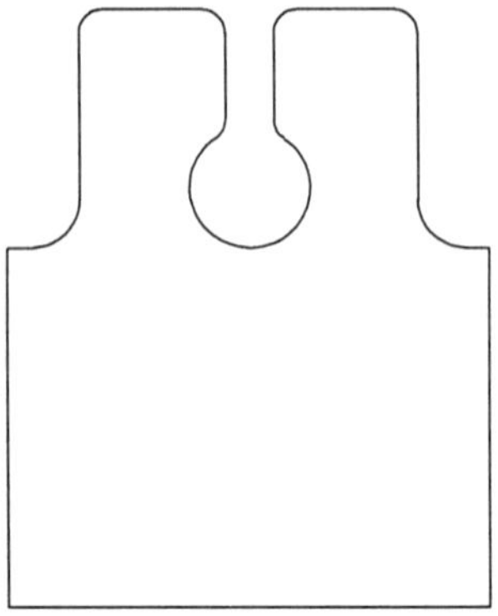

Figure 9:2 Small Punches

1. Footed (Figure 9:3)

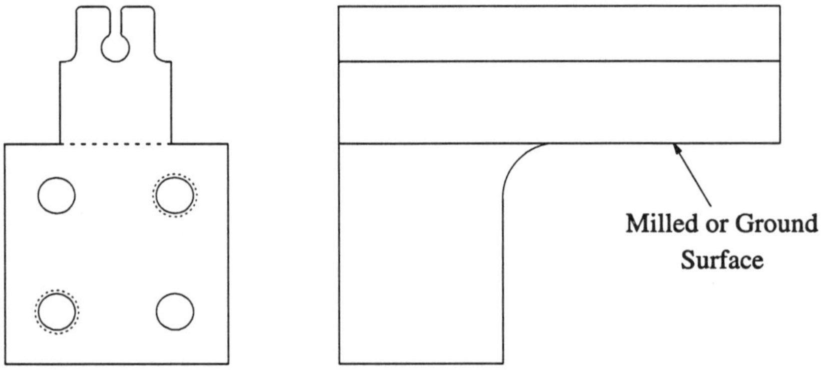

Figure 9:3 Footed Punch

2. Shoulder (Figure 9:4)

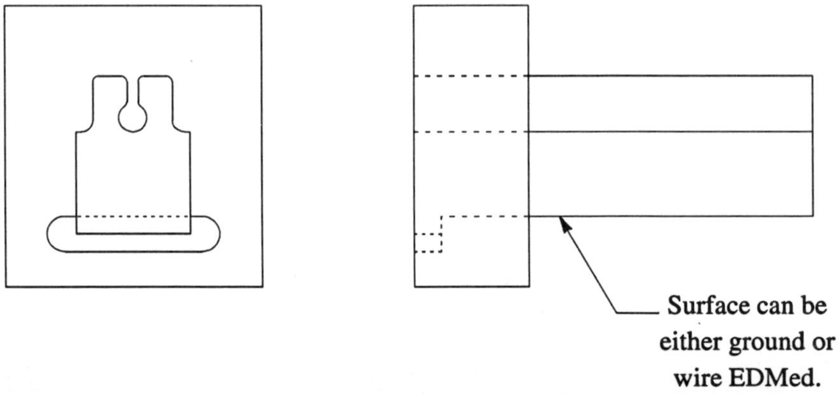

Figure 9:4 Shoulder Punch

3. Keyed In--Toe Clamps (Figure 9:5,6)

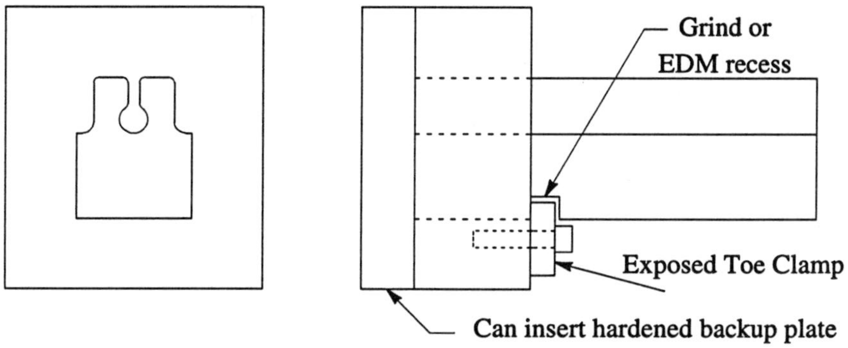

Figure 9:5 Keyed In-Toe Clamp

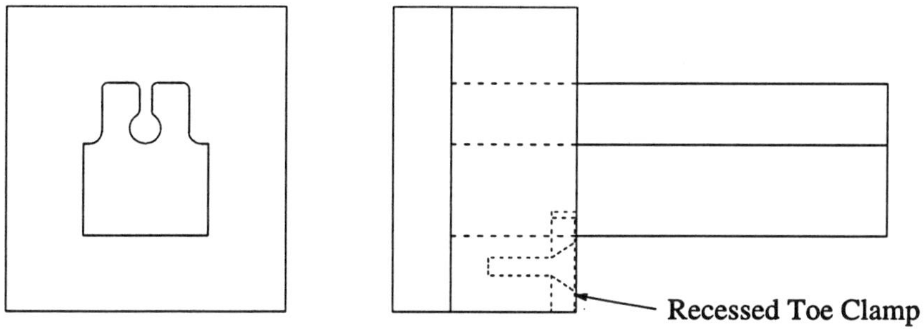

Figure 9:6 Recessed Toe Clamp

4. Keyed In--Keyway (Figure 9:7)

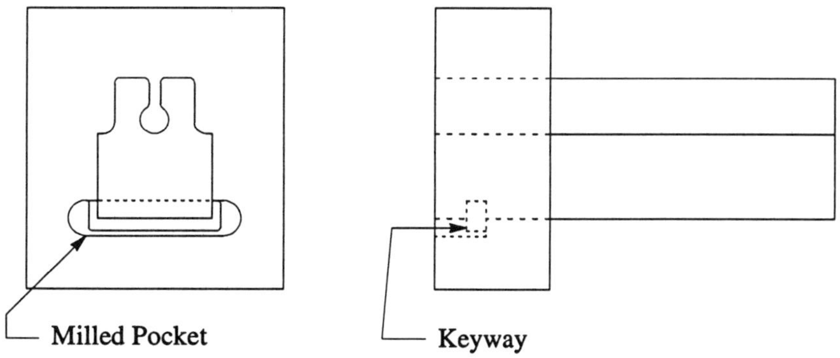

Figure 9:7 Keyway Insert

5. Press Fit (Figure 9:8)

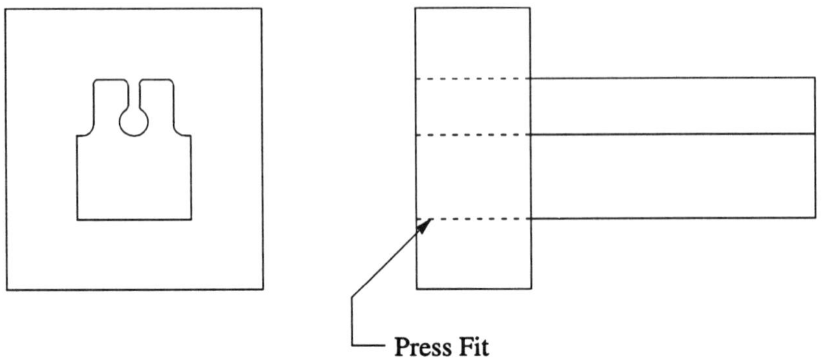

Figure 9:8 Press Fit
Not recommended for heavy stripping pressures.

6. Peened Edge (Figure 9:9)

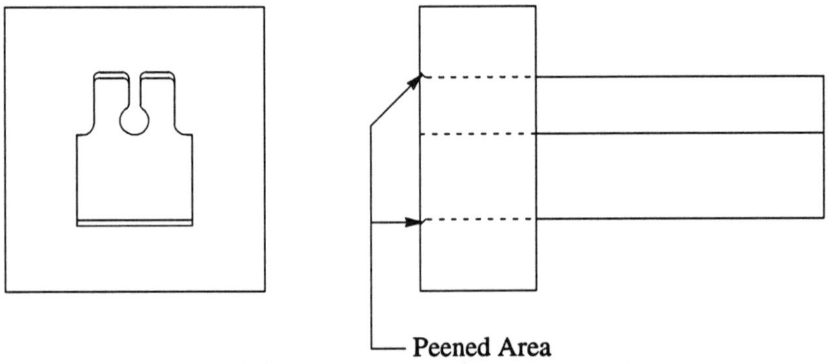

Figure 9:9 Peened Edge
Not recommended for heavy stripping pressures.

7. Dowel Pin Reamed (Figure 9:10)

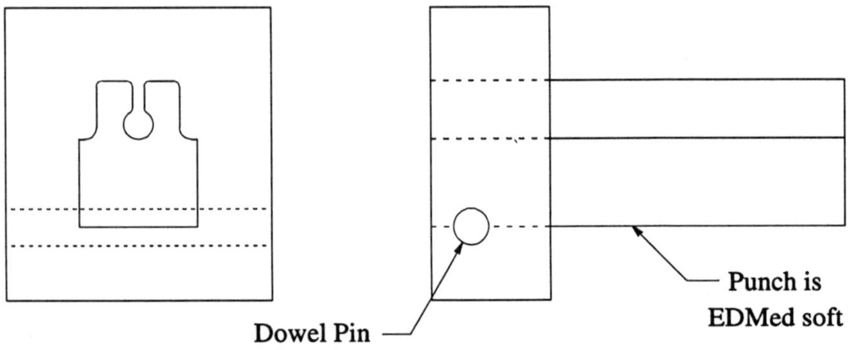

Figure 9:10 Dowel Pin
Punch is EDMed before heat-treating.

8. Dowel Pin EDMed (Figure 9:11)

Should the pressed fit or the peened fit ever come loose, a dowel hole can be EDMed.

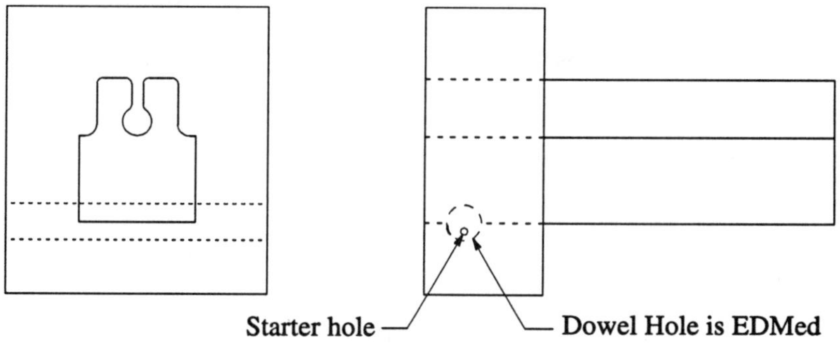

Figure 9:11 Dowel Pin EDMed

9. Set Screws (Figure 9:12)

Should the pressed fit or the peened fit ever come loose, a slot can be ground and set screws used.

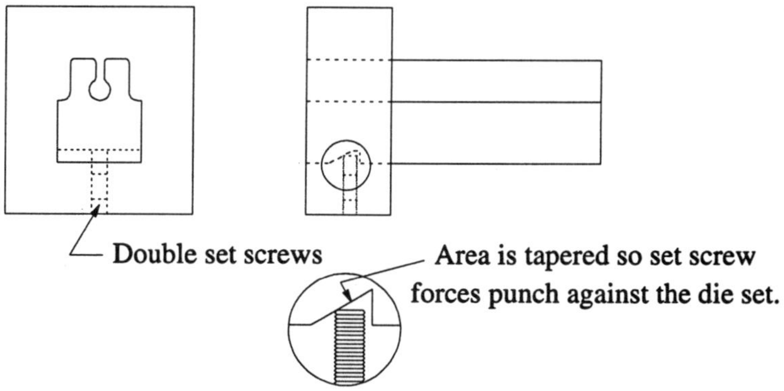

Figure 9:12 Set Screws

10. Socket Head Cap Screw (Figure 9:13)

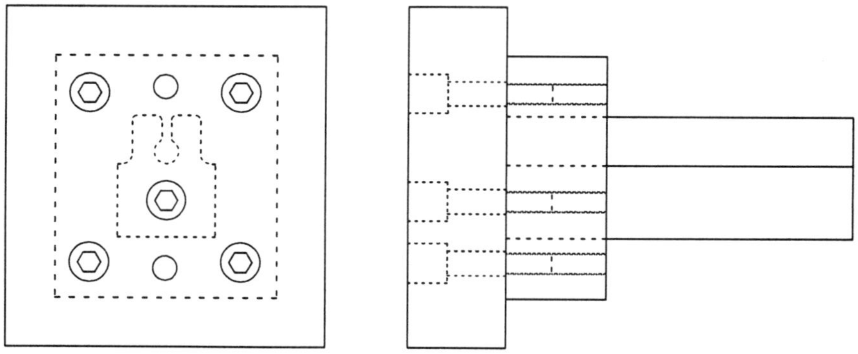

Figure 9:13 Socket Head Cap Screw

EDM punch holder and hold punch with a socket head cap screw.

11. Ball Bearings (Figures 9:14,15)

Use a carbide ball end mill to put a radius indentation in punch. Use the same end mill to the same depth as the punch to mill out the sides of the punch retainer. Put in hardened steel bearings to hold punch.

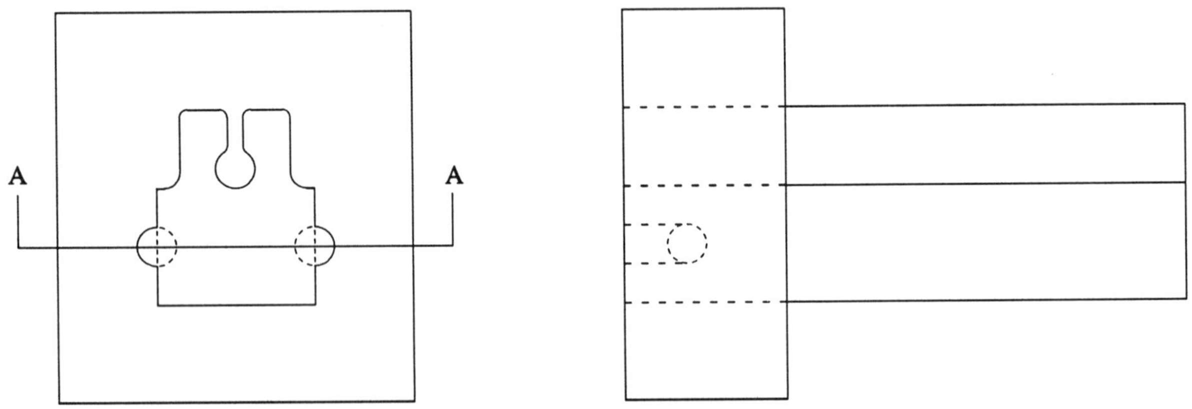

Figure 9:14 Ball Bearings

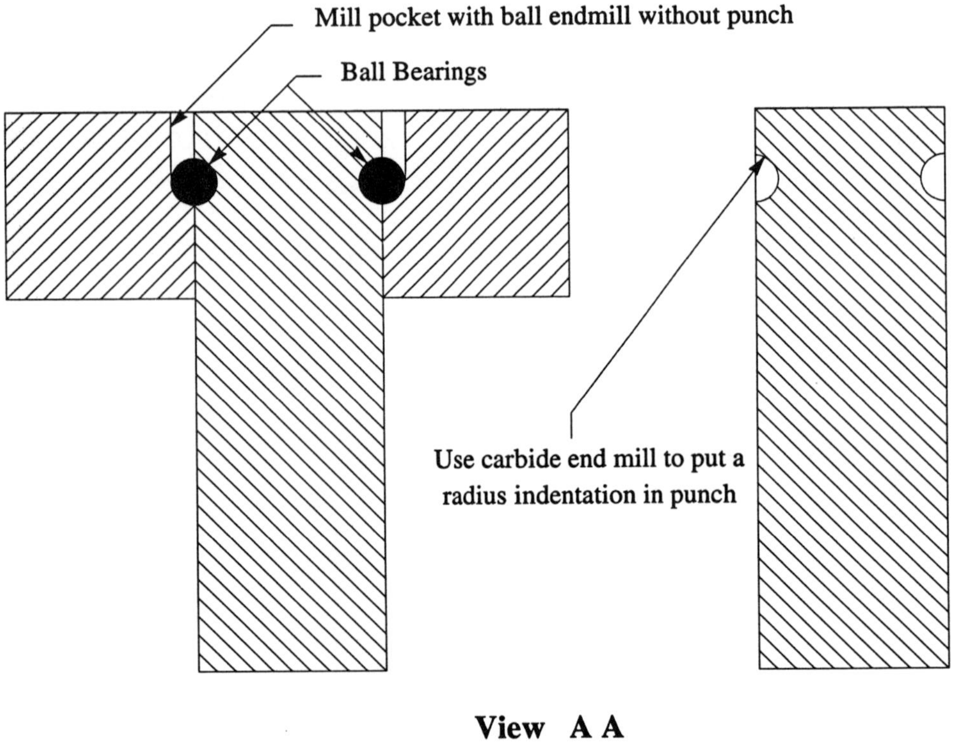

Figure 9:15 Ball Bearing Construction

C. Skim Cutting

On close tolerance dies, skim cuts are made depending on the accuracy of the punch and die sections. The die sections cause no difficulty in skim cutting, since the cavity is open. However, in skim cutting the punch has to be held with a tab. The tab is made in a straight section, and then cut off in the final cut and ground to size. See Figure 9:16.

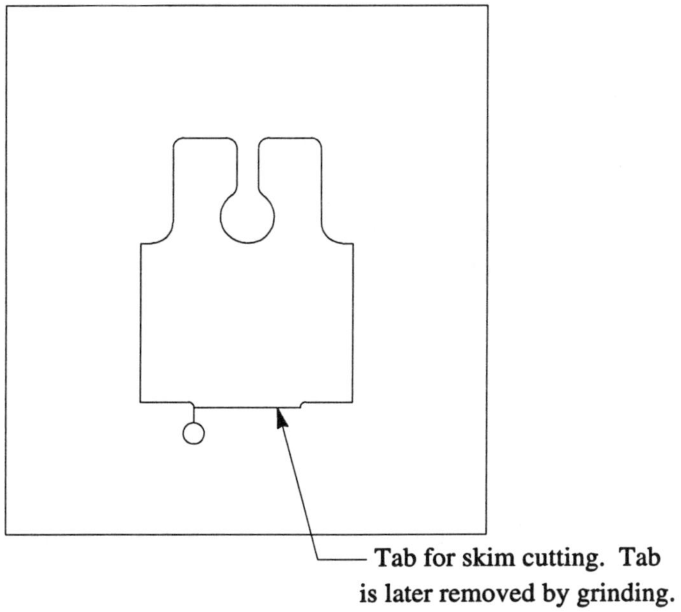

Figure 9:16 Skim Cutting Punches for Precision Dies

Die Sections

A. Heavy Blanking Dies

Always use a sufficiently large die block to avoid splitting when thick steel is being cut. The nominal cost is well worth not having to remake the die.

B. Avoid Sharp Corners

Sharp corners are the weakest area of a die section. When possible, avoid them.

C. Heat-Treating

Cut die sections in the heat-treated condition. This avoids heat treat distortions.

D. Large Die Sections

Large sections should be double and even triple tempered in order to remove all the stresses, particularly in close tolerance dies. Even then some stresses may remain. To ensure minimum stresses on precision dies, cut out the mid-section on a band saw, leaving at least a 1/4 inch of wall, or put in relieving slots. See Chapter 4, "Cutting Large Sections."

E. Tapers

1. Taper and Land

Due to the accuracy of wire EDM, there is no need for large tapers. Dies can be made with a taper and land; however, most die sections can be tapered right up to the top of the die section. See Figure 9:17. With a 1/4° taper, the die will only become .001 larger per side when .250 is removed.

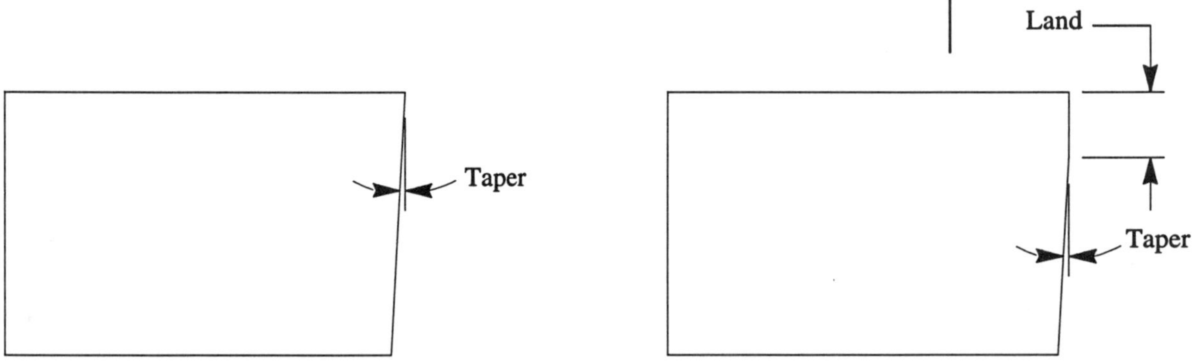

Figure 9:17 Taper and Land

2. Straight Cut and Taper

Wire EDM can go from taper to straight, as shown in Figure 9:18. In areas where the punch can be supported in the die, stipulate a straight land to support the heel of the punch.

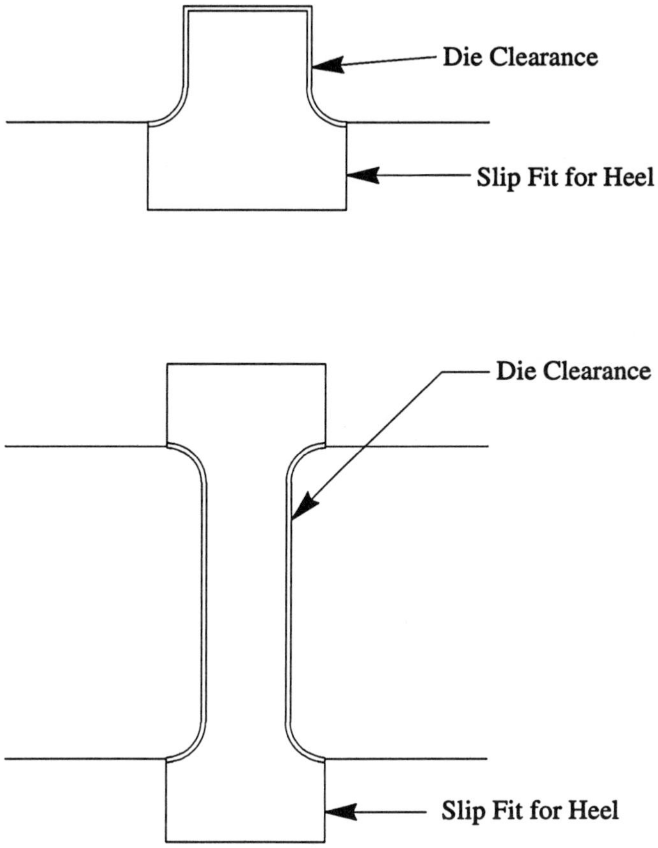

Figure 9:18 Straight Cut and Taper

Wire EDM has provided the tool and die designer with many options in building dies. The next chapter will show one of the fastest and most cost effective ways to produce dies.

10

One-Piece Stamping Dies

10 One-Piece Stamping Dies

Blanking Die

Wire EDM has made it possible to produce a high quality inexpensive die from one piece of tool steel. Following is a description outlining this method.

A. Desired Stamping

The desired stamping can be either the slug or the blank. See Figure 10:1.

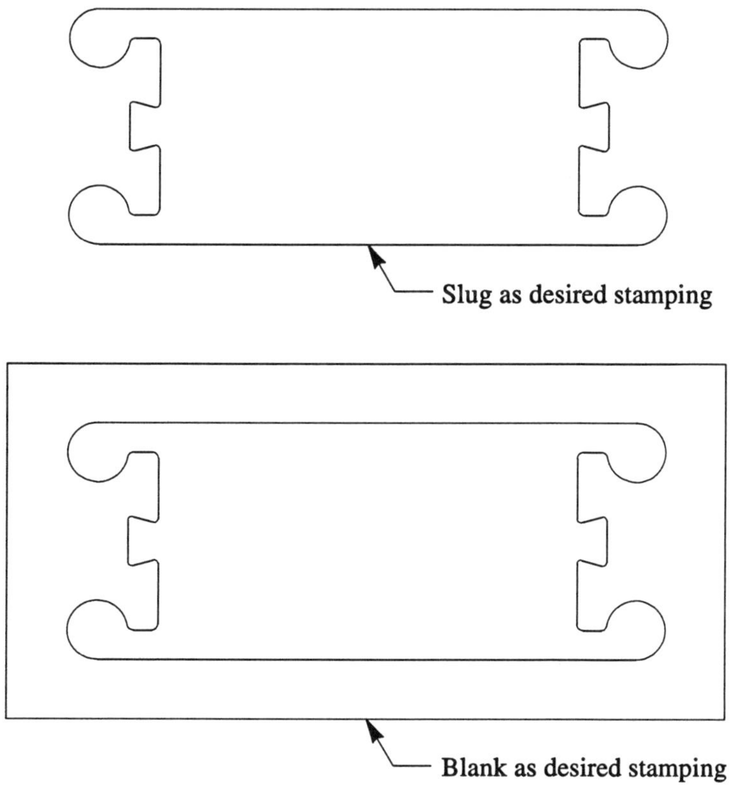

Figure 10:1 Desired Stamping: Slug or Blank

B. Preparing the Tool Steel Blank

Drill, ream and tap all holes for punch, die and stripper as shown in Figure 10:2.

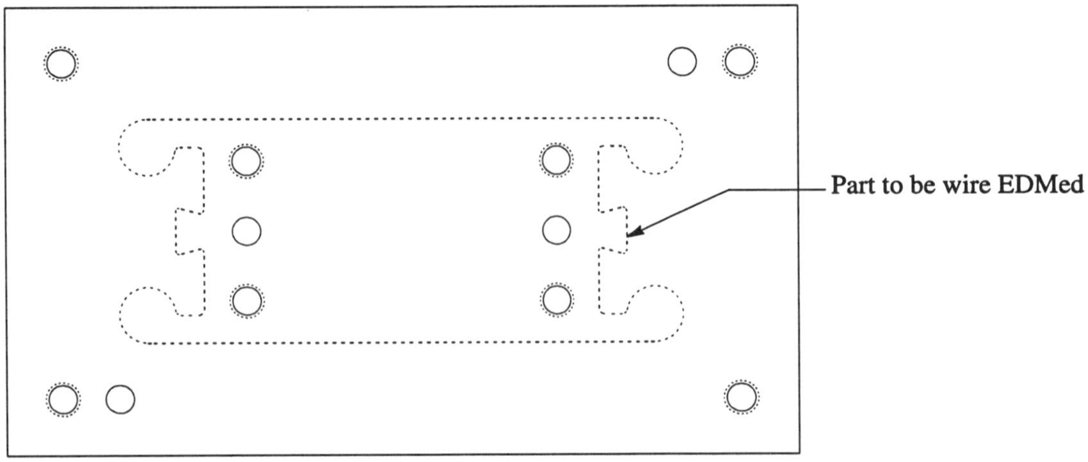

Figure 10:2 Part to be Wire EDMed
Drill, ream and tap all desired holes.

C. Placement of Starter Hole

The starter hole can be placed in either the punch section or the die section. The small line created with wire EDM is in most cases negligible; however, the line should be placed in the part which will produce the scrap.

1. Punch Shape is the Desired Stamping

If the desired stamping will be the shape of the punch, then the starter hole should be in that punch. The part will take the shape of the die section. Put the starter hole about 1/4" from the cutting edge.

Rule: Starter holes should always be placed in the desired shape that is in the die. Ex: If the punch is the desired shape, the starter hole should be put in the punch section of the die. See Figures 10:3,4.

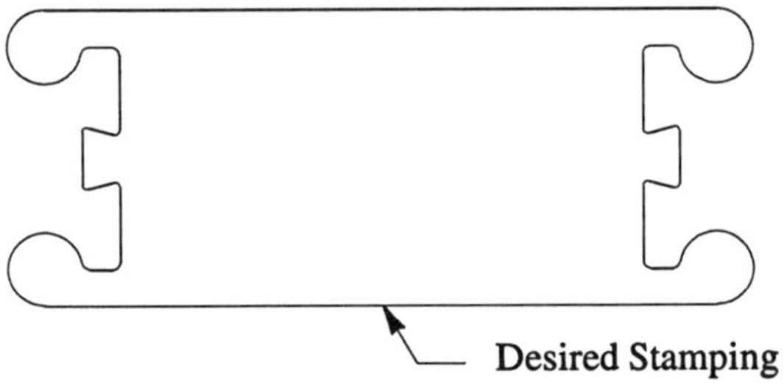

Figure 10:3 Desired Stamping is the slug

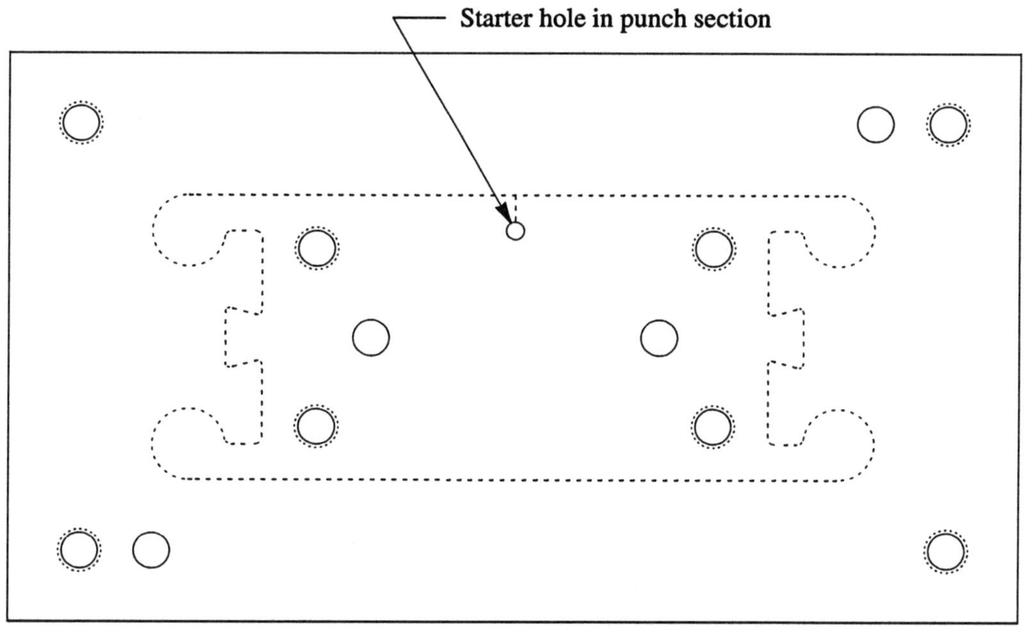

Figure 10:4 Placement of Starter Hole

If punch is desired shape, put starter hole in the punch section of the die.

2. Blank Shape is the Desired Shape

If the desired stamping will be in the shape of the remaining blank, then the starter hole should be placed in the die section. The part will take the shape of the punch. See Figure 10:5,6.

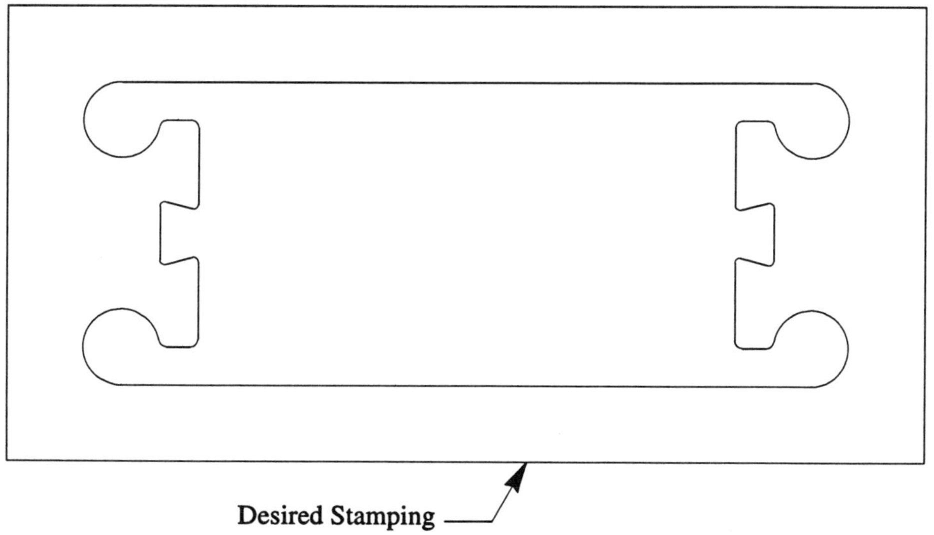

Figure 10:5 Desired Stamping is the Blank

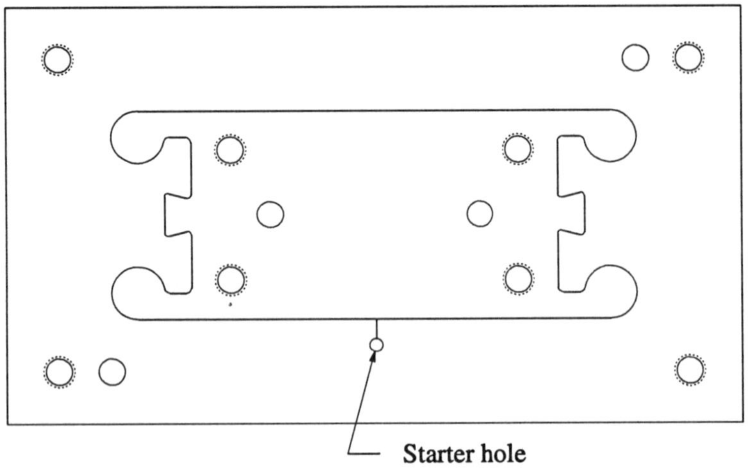

Figure 10:6 Placement of Starter Hole
If the remaining blank is the desired shape, put the starter hole in blank section of the die.

D. Harden the Tool Steel Blank

After all holes have been put in, the tool steel blank should be heat-treated and tempered to desired hardness. In close fitting dies, the steel should be stress relieved and double or triple tempered. Air hardening tool steels should be used. Oil hardening steels have more internal stress after heat-treating.

E. Stripper Plate

Transfer all holes into the stripper plate, including the starter hole. Remove the stripper plate.

F. Punch Holder

Transfer holes from punch to the punch holder, as illustrated in Figure 10:7. Drill and ream all punch holes. Remove punch holder.

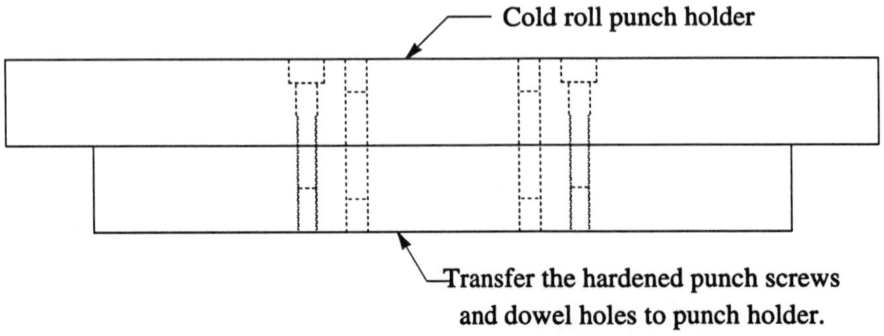

Figure 10:7 Mount Hardened Punch Section
The punch section is mounted to the punch holder with screws and dowels.

G. Mount Die Blank on Die Set

Drill and ream the die block to the bottom die set as in Figure 10:8. Bolt and dowel the die blank to the bottom die set. Do not remove the bolted die section from the die set.

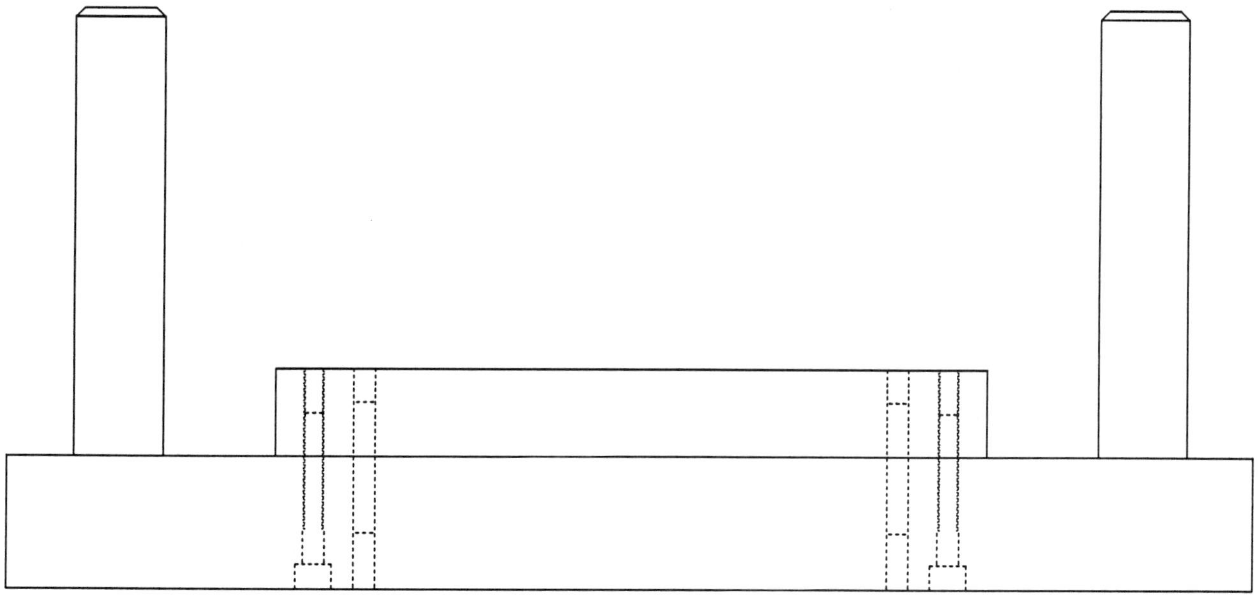

Figure 10:8 Mount Hardened Die Section
The die section is mounted on the die set with screws and dowels.

H. Mount Punch Holder on Die Set

Bolt and dowel punch mounting plate to punch section that was previously done. Put on upper die shoe and drill and tap for bolts, then drill and ream for dowel pins into the punch mounting plate.

Mounting the punch holder before the die is wire cut eliminates the need for the difficult task of lining up the punch with the die sections. This method produces a perfect alignment as illustrated in Figure 10:9.

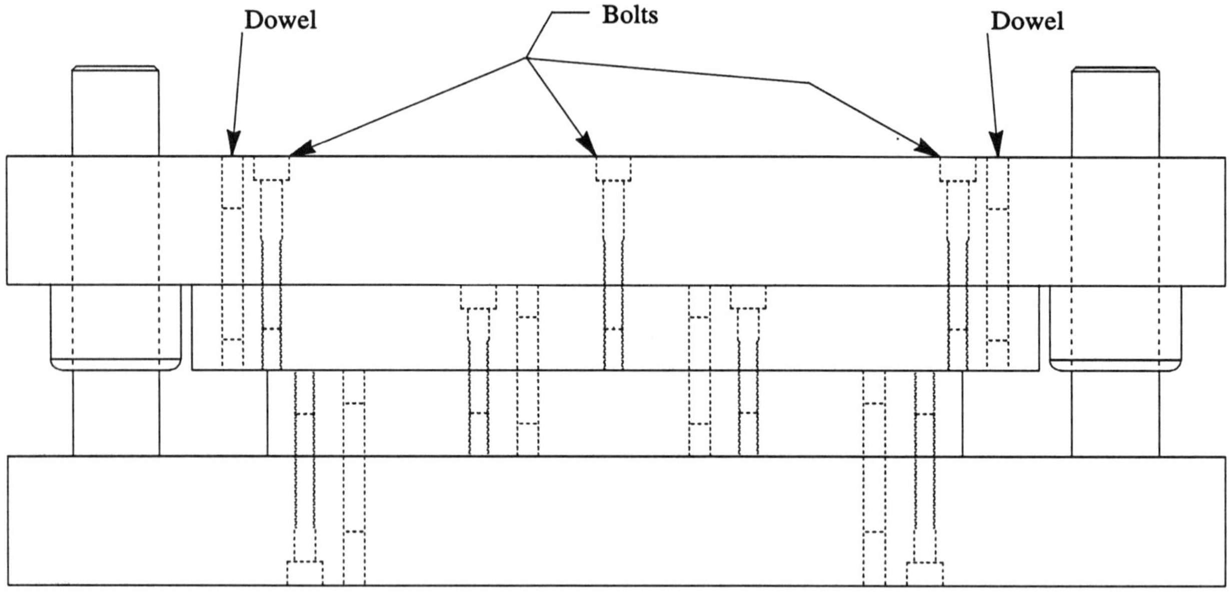

Figure 10:9 Mount Die Section on Upper Die Set

With the die section mounted on the die set, and the punch holder mounted on the hardened blank, the cold roll punch holder is screwed and bolted and reamed in place. Now when the die is wire EDMed, there will be a perfect alignment.

I. Mount Stripper on Bottom of Die Section

By mounting the stripper on the bottom of the die section, it will be cut at the same time as the regular die section and have proper clearance all around. The dowel pins can be used to hold the stripper. See Figure 10:10.

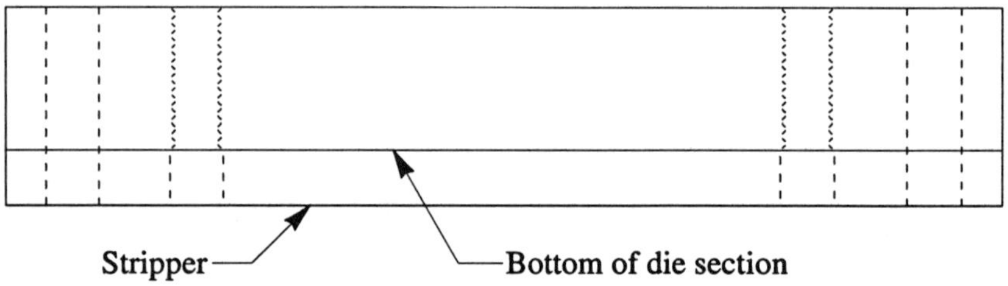

Figure 10:10 Wire EDM Stripper with Die Section

J. Wire Cut Punch, Die and Stripper

The EDM programmer calculates the exact taper needed to produce the proper clearance. With one cut, the punch, die and stripper will be produced as in Figure 10:11. The stripper slug can be used to extend the punch. Land can be easily done with a skim cut. See Figure 10:12.

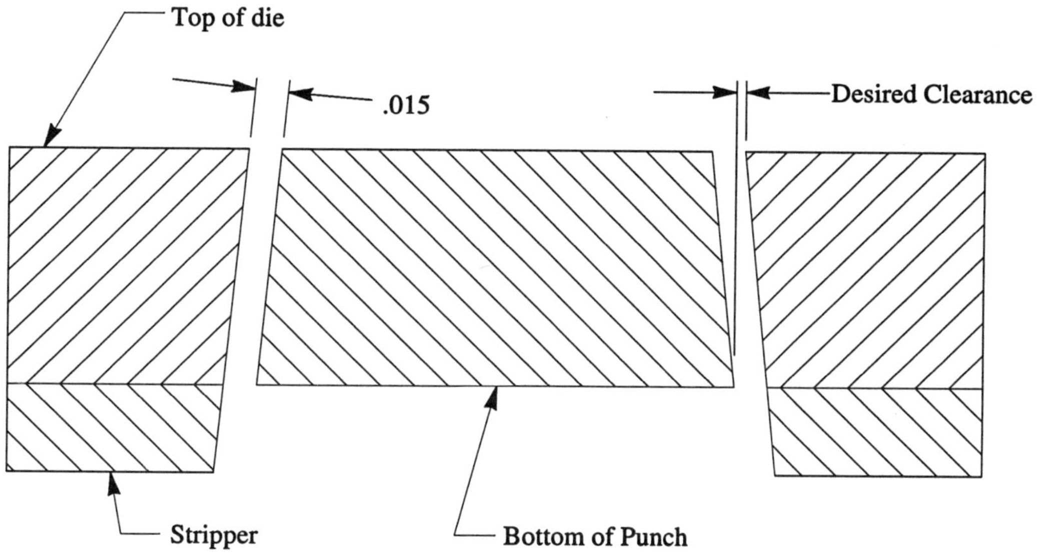

Figure 10:11 Punch, Die & Stripper Can Be Made with One Cut

Example: 14 gage cold roll steel with 12% clearance per side.
EDM programmer will determine the proper angle to cut the die section.

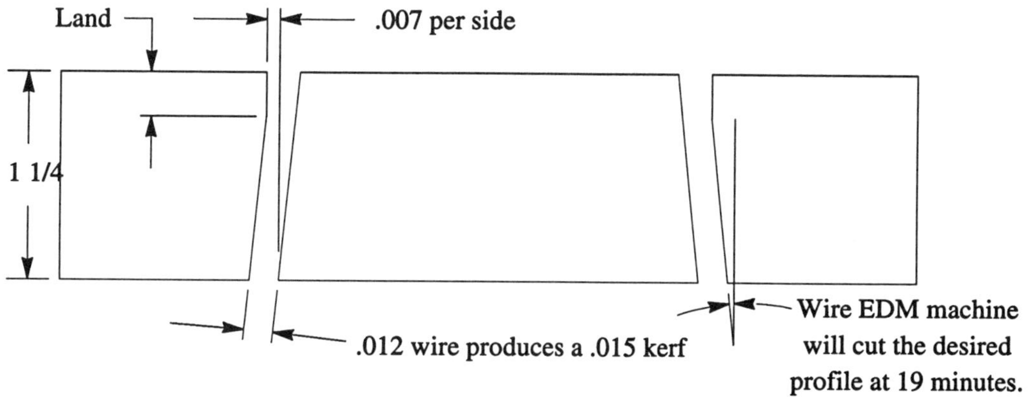

Figure 10:12 Calculating Desired Clearance

(Clearance is exaggerated)

Total Burr-Free One-Piece Blanking Die

For most dies, placing the starter outside of the punch or die and leaving the narrow kerf for cutting has a negligent effect on the part. On thin materials below 1/32" thick, there can be a raised area that for precision stamping parts may be objectionable.

There is another method in making a one-piece die where there is no kerf at all. The starter hole is drilled on an angle where it intersects directly in the middle of the punch and die section. See Figure 10:13 & 10:14.

If there are other punch and die holes in part, the starter hole should be precisely located. In the other method of using a straight starter hole and leaving a kerf, the placement of the starter hole is not critical.

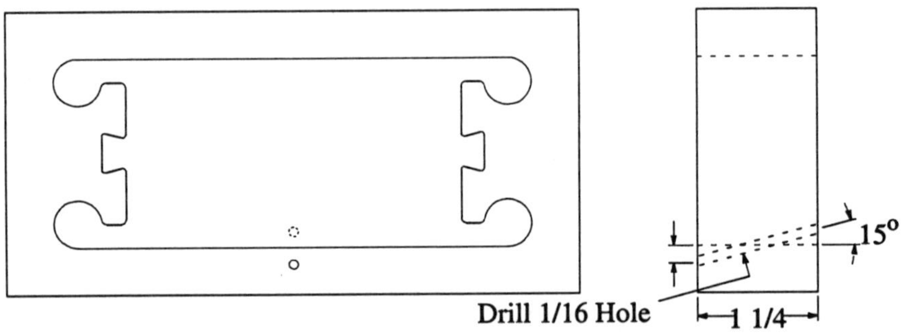

Figure 10:13 Burr-Free One-Piece Blanking Die
Recommended for precision stampings.

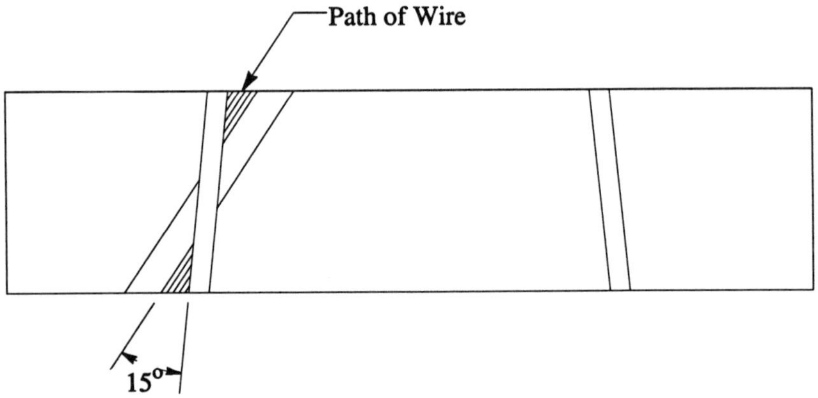

Figure 10:14 Path of Wire
The path of the wire will leave a kerf halfway on the top of the punch and halfway on the bottom of the die where stamping does not take place.

Compound Blanking Dies

A. Desired Stamping (Figure 10:15)

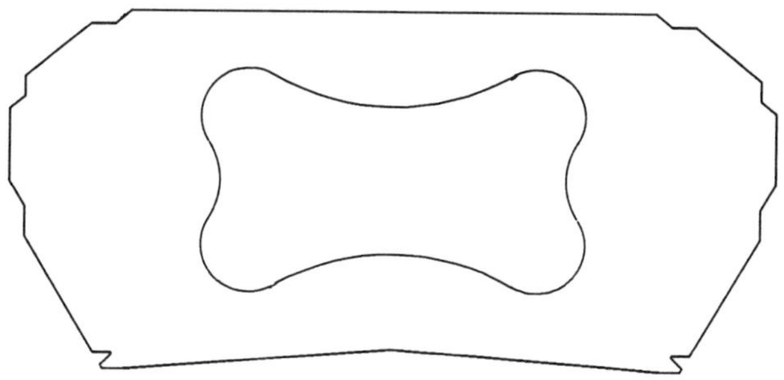

Figure 10:15 Desired Stamping

B. Prepare Tool Steel Blank

Drill, tap and ream all necessary holes, including starter holes. See Figure 10:16. Remember: Put starter holes in the desired shape that is in the die. Harden and temper.

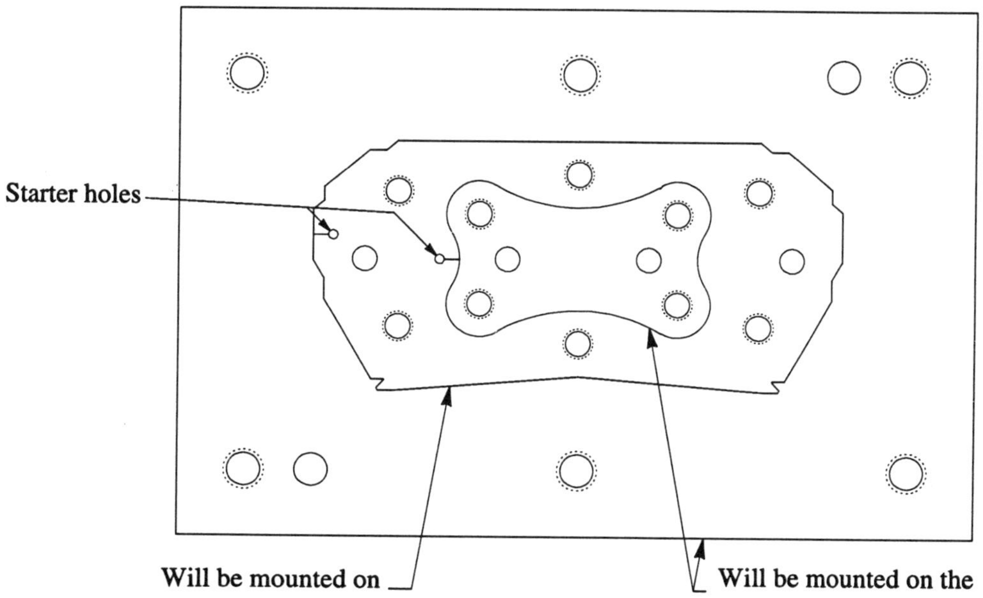

Figure 10:16 Prepare Tool Steel Die Section
Put in all desired holes before hardening.

C. Mount Punch Holder

Mount punch onto a punch holder as in Figure 10:17. Make sure the punch holder is large enough to hold the stripper bolts and springs. Remove punch holder.

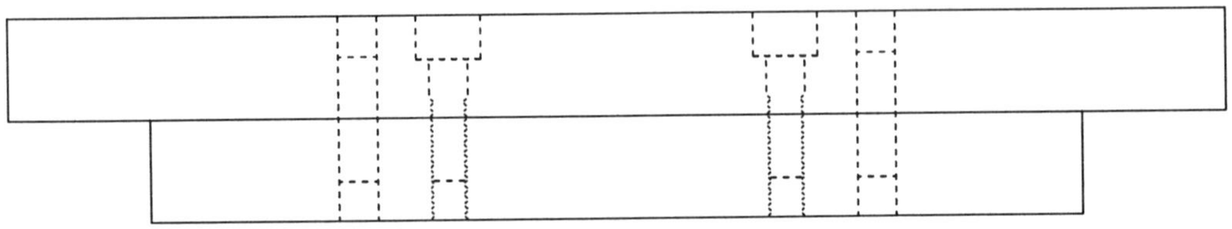

Figure 10:17 Mount Punch on Punch Holder
(Mount middle portion of the die.)

D. Mount Die Block on Bottom of Die Set

Drill and ream all holes for sections that will be mounted on the bottom of the die section as shown in Figure 10:18.

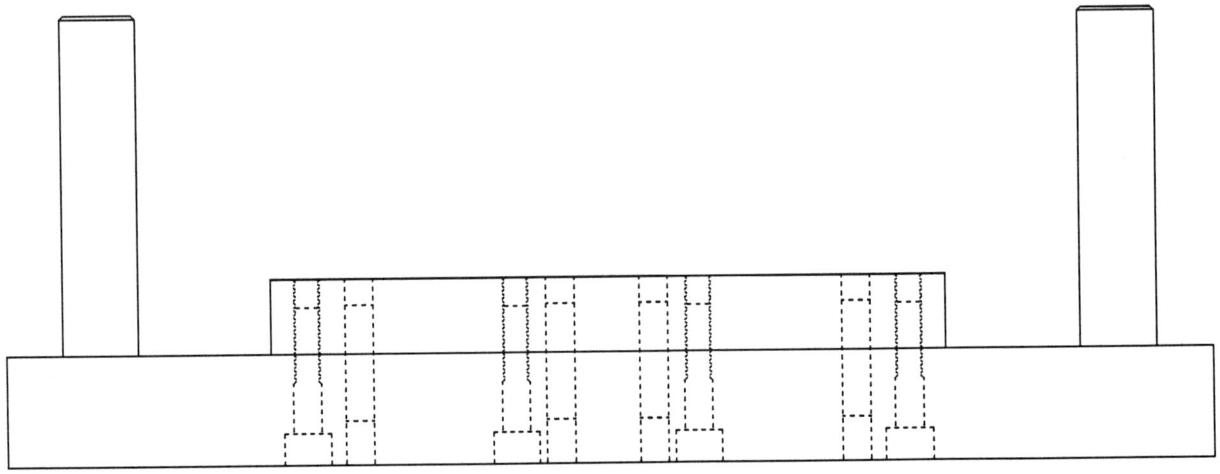

Figure 10:18 Mount Die Blank on Bottom Die Set
(Mount the inside and outside sections of the die.)

E. Mount Punch Holder on Top of Die Set

Mount punch on the punch holder, as illustrated in Figure 10:19, with dowel pins. Drill, ream and tap holes from top of die set to punch holder. Now the die section can be removed for wire EDMing. Mounting the punch holder before wire EDMing creates a perfect alignment.

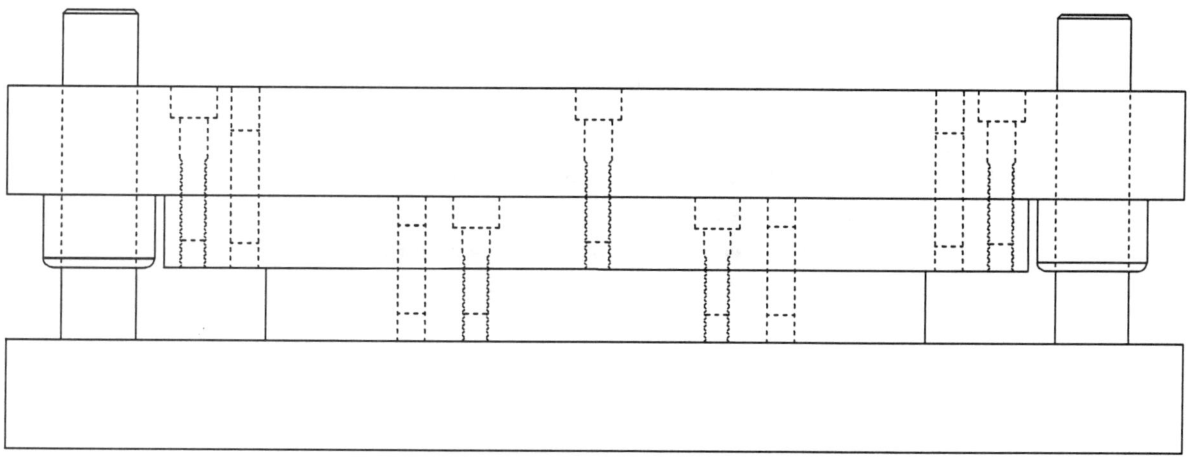

Figure 10:19 Mount Punch Holder
Mount punch on punch holder, then bolt and dowel punch holder to upper die set. This procedure will guarantee a perfect alignment after the die is EDMed.

F. Stripper Plate

Since this is a compound blanking die, parts of the stripper will be on both the top and the bottom of the die shoe. Drill no holes on the stripper except for the two starter holes.

If the angle of the cut is relatively straight, then the stripper can be clamped on the die section and cut at the same time. Otherwise, the stripper may have to be cut separately.

G. Wire EDM Compound Die

The EDM programmer will calculate the exact angle for proper clearance. From one piece of tool steel, a high performance inexpensive compound die is produced. See Figure 10:20.

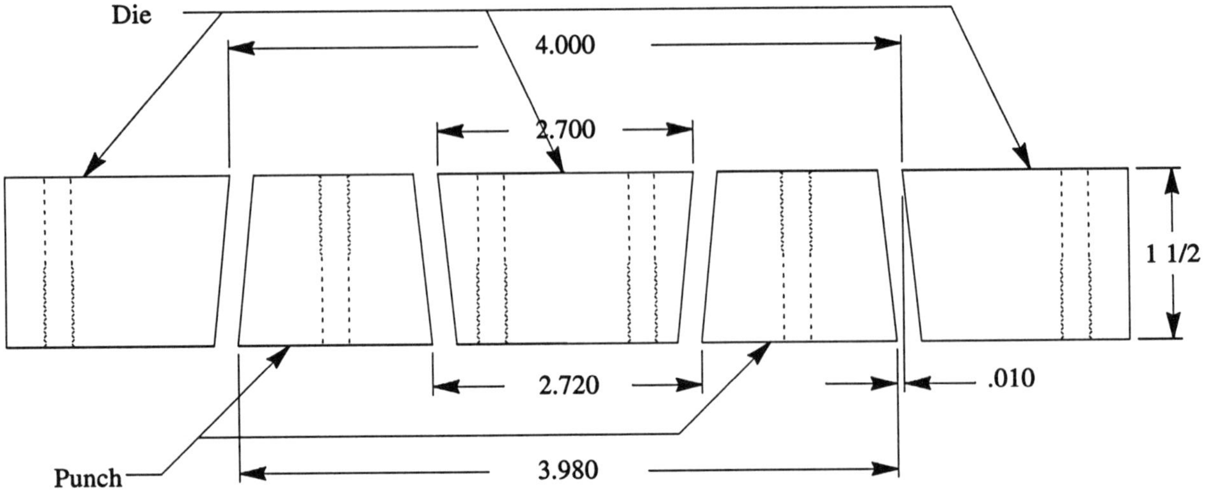

Figure 10:20 Compound Die from One Piece of Tool Steel
From one piece of tool steel, a high performance inexpensive compound die can be produced. (Clearance is exaggerated.)

H. Completed Die

1. Stripper with Springs

Mount stripper with stripper bolts and springs on both top and bottom of die set. See Figure 10:21.

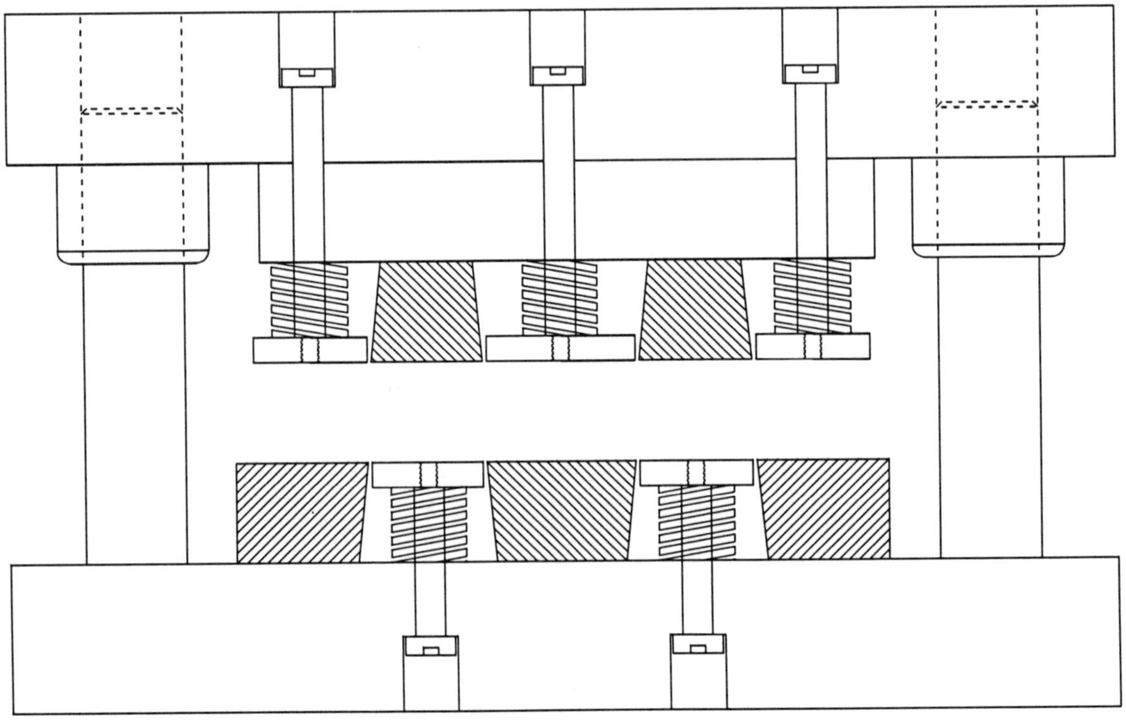

Figure 10:21 Completed Compound Die with Spring Mounted Strippers

2. Stripper with Knockout

The advantage of a knockout die is that the scrap and part will be separated. See Figure 10:22.

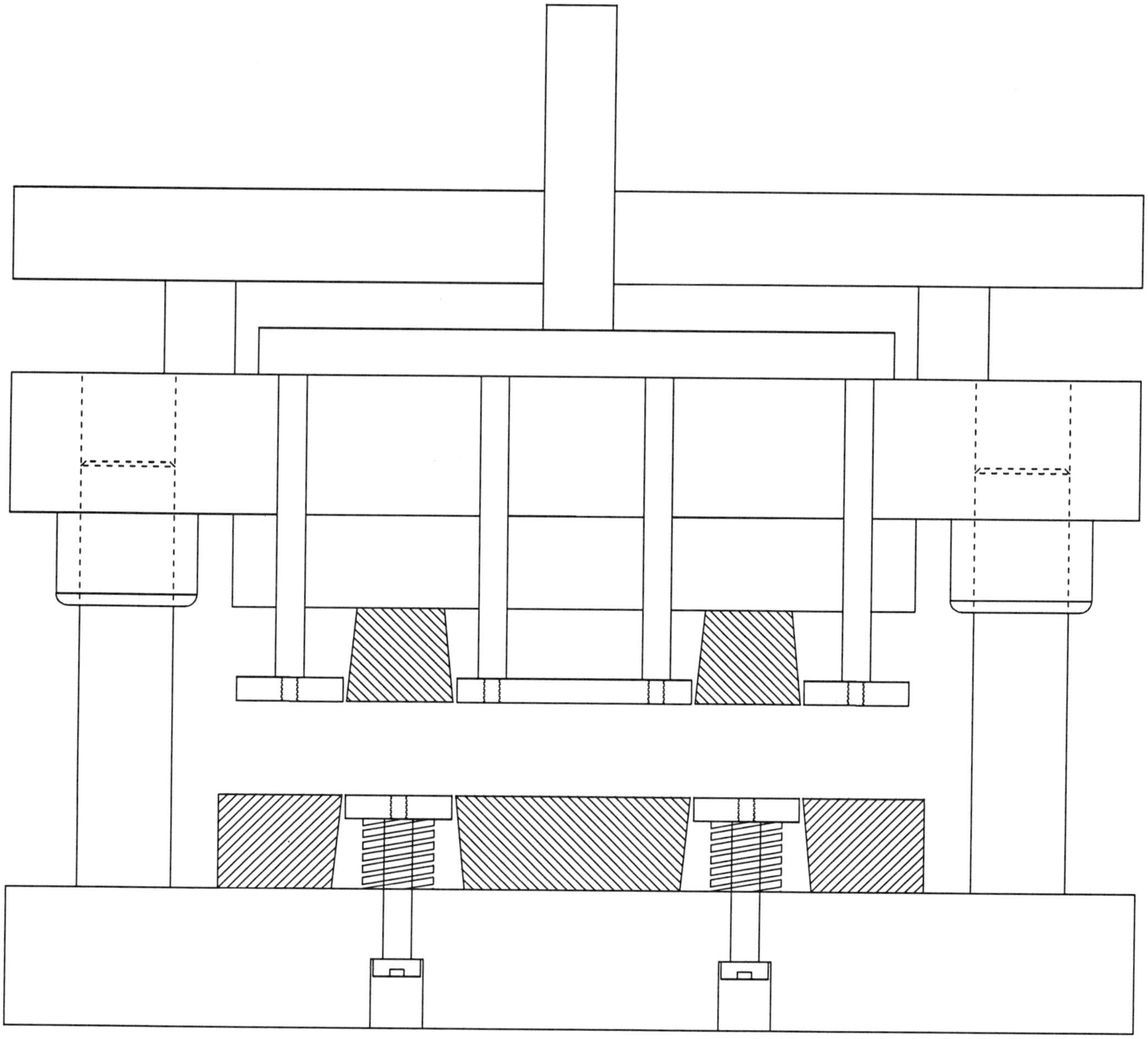

Figure 10:22 Completed Die with Knockout
Knockout removes slug from part.

Appendix

(Questions for Textbook Use)

Chapter 1
Fundamentals of Wire EDM

1. What is one of the most important issues that managers need to understand in order to become and stay competitive?

2. List the decades of wire EDM machines and their relative cutting speeds.

3. Why should wire EDM no longer be considered non-traditional machining?

4. What is an added benefit of using the wire EDM process over conventional machining?

5. Because of wire EDM, what are today's design engineers doing?

6. List the two types of EDM.
   A.

   B.

7. What is the distinct advantage of using a shaped electrode instead of using wire EDM?

8. Though wire EDM is the same as any other machining tool, what is unique about it?

9. Describe in detail how wire EDM works. Include in your description: electrical pulses, wire electrode, ionized water, dielectric, voltage, vaporize, pressurized cooling fluid, filter, suspended solids, resin, electrical conductivity, chiller, and servo system.

10. What accounts for the extreme accuracy of wire EDM?

11. What is one of the best descriptions of wire EDM?

12. Describe the following three types of wire EDM:
   A. Two axis

   B. Simultaneous four axis

   C. Independent four axis

Chapter 2
Profiting With Wire EDM

1. What is one of the added benefits of using wire EDM which does not produce additional cost to the user?

2. Why can extremely thin pieces be machined with wire EDM?

3. What accuracies and finishes can be produced with wire EDM?

4. Draw a picture of a part produced by stamping and a part produced by wire EDM.

5. Give some practical illustrations of how using wire EDM can be used to repair worn or broken parts.

6. What effect does material hardness have on wire EDM?

7. Describe digitizing.

8. Describe air machining.

9. Why is the wire EDM process so reliable?

10. List the twenty-three parts that can be produced with wire EDM. Can you name any other parts?

11. What is the advantage of cutting hardened collets with wire EDM rather than with conventional tooling?

12. List the three advantages of wire EDM over plunge EDM.

13. What is the difference between laser (Light Amplification by Stimulated Emission of Radiation) and wire EDM?

14. What is the advantage of using laser over wire EDM?

15. What is the advantage of using wire EDM over laser?

16. What are the basic kinds of materials that can be cut with wire EDM?

17. How are machining costs usually determined?

Chapter 3
Proper Procedures

1. List the three methods to pick up dimensions.

2. By using an edge and two holes, what can be done with wire EDM?

3. To ensure accuracy, how should the edges be prepared?

4. What is used if the part has unfinished edges?

5. Why is scale a problem with wire EDM?

6. What can be done to avoid scale?

7. If there are obstructions, what can be used?

8. What happens if a pickup hole is not square?

9. What is the most accurate method to pick up dimensions for accurate parts when using holes?

10. Where is the best place to put starter holes?

11. On thin cuts, where should the starter hole be placed? Why?

Chapter 4
Understanding the Wire EDM Process

1. What tolerances can wire EDM machines hold?

2. What kind of finishes can these machines produce?

3. Why is there such an excellent finish with wire EDM when compared with laser or water jet cutting?

4. What is wire kerf?

5. What is the smallest corner and inside radius a .012 wire can produce?

6. List the four factors for skim cutting.

7. What determines the cutting speeds of material?

8. What will wire EDM do if there are impurities in the metal?

9. What causes metal to move during wire EDM?

10. What can be done to avoid the problem of metal moving?

11. What is the purpose of leaving a frame around a part being cut with wire EDM?

Chapter 5
Reducing Costs

1. Why is it important to try to create one slug when cutting with wire EDM?

2. What happens when cutting close to an edge with wire EDM?

3. What is the most efficient method of cutting with wire EDM? Why?

4. Why is it sometimes more cost effective to machine parts after the EDM process?

5. Why can holes in stacked parts cause problems when cutting with wire EDM?

6. If stacked parts need holes or other machining, what is the best method for wire EDM?

7. What precautions need to be adhered to when stacking parts for wire EDM concerning non-conductive materials?

8. In precision production parts, why is it important for parts to be machined the same and square?

Chapter 6
Production Wire EDM

1. What makes wire EDM so much more cost effective today?

2. What is one of the major problems with designers and engineers concerning wire EDM?

3. List the advantages of job shop wire EDM companies?

4. Explain why manufactures versus job shops are generally not as efficient in their operations.

5. Calculate the yearly cost of owning a wire EDM machine. Use approximate cost figures.

6. Describe the wave of the future.

Chapter 7
Various Wire EDM Applications

1. What will happen as the capabilities of EDM are discovered?

2. Though machines are designed to cut certain heights, what can be done to change them?

3. List three unique applications for wire EDM (Ex. Making precision components for surgical instruments).

A.

B.

C.

Chapter 8
Advantages of Wire EDM for Die Making

1. Describe the history of tool and die making from using hand files and filing machines, to precision form grinding, and wire EDM.

2. What role has changed for skilled tool and die makers?

3. What are the advantages of solid dies over sectionalized ones?

4. What are the advantages of using wire EDM concerning spare parts for stamping dies?

5. How can broken dies be salvaged with wire EDM?

6. Why does the fine textured finish prolong die life?

7. What can be done for high wear areas in dies?

Chapter 9
Wire EDMing Punch and Die Sections

1. On large punches, how can they be held in the die set?

2. List the eleven methods by which small punches can be held.

3. What is the weakest area of a die section?

4. What precautions should be taken when heat-treating?

5. What is the advantage of going from a taper cut to a straight cut in die making?

Chapter 10
One Piece Stamping Dies

1. What is the rule for starter holes?

2. What is the advantage of mounting the hardened die and punch section to the die before cutting with wire?

3. What can be done with the stripper plate when cutting the tool steel?

4. Describe a compound blanking die.

5. Describe the difference between a stripper with springs and a stripper with a knockout.

Index

Index

A

Accuracy 36, 42, 67
  Inspection Equipment 96
  Leaving A Frame 76
Air Machining 41, 83
Aluminum 21, 52, 73
Appendix 147

B

Barreling Effect
  See Wire
Beryllium Copper 52
Brass 52
Burr Free 37

C

Carbide 21, 46, 52, 72 --73
  Ferro-Tic (Machinable Carbide) 52
Chiller 24
Clamping 77
Collets 43
Compound Blanking Dies
  See Dies
Conventional EDM
  See Plunge EDM
Copper 52
Corners
  See Wire
Costs 52
  Machine Costs 95
CPM 10V
  Steel 52
Cutting Speeds
  See Speeds

D

Deionized Water 23
Dielectric Fluid 23
Dies 109
  Advantages of Wire EDM Dies 112
  Burr Free One Piece Blanking Die 138
  Compound Blanking Dies 139
  Large Die Sections 126
  Large Punches 119
  One Piece Stamping Dies 131
    See also Punches
  Small Punches 120
  Starter Hole 132
  Straight Cut and Taper 126
  Taper and Land 126
Digitizing 40
Dowel Holes 113

E

Edge Preparation 58

F

Ferro-Tic
  See Carbide
Finishes 36, 67, 721
Flushing 72, 81, 83
Frame 76
  Strength of Frame 76

G

Gages 42
Gears 47
Graphite 41, 52

H

Hardness 38, 74
Hastalloy 52

I

Impurities 73
Inserts 38, 113-114
Inspection 94
Internal Splines 47

J

Job Shops
  Advantages 93

K

Kerf
  See Wire
Keyways 43

L

Lasers 50
Layout 63

M

Machine Speed
    New Machines 20
    Old Machines 20
Molds 45

N

Nitronic 52
Nontraditional Machining 20
Notch Test Specimens 47

O

One Piece Stamping Dies
    See Dies

P

Parting 44
Pick Up 57, 60
Plunge EDM 22
    Disadvantages 49
Power Supply 25
Production EDM 21, 35, 46, 86, 91
Prototypes 37
Punches
    Ball Bearing 124
    Dowel Pin EDMed 123
    Dowel Pin Reamed 122
    Footed 120
    Keyed In 121
    Large 119
    Peened Edge 122
    Press Fit 122
    Set Screws 123
    Shoulder 120
    Small 120
    Socket Head Cap Screw 123

Q

Questions 147

R

Radii
    Wire 69
Ram EDM
    See Plunge EDM
Recast 74

S

Scale 59, 86
Servo 24
Short Run Stamping 45
Sink EDM
    See Plunge EDM
Skim Cutting 69, 71, 74, 125
Speeds 81, 83
Squaring Block 59
Stainless Steel 52
Starter Holes 60
    Dies 132
    Placement of Starter Holes 62
    Relieved Holes 61
    Unsquare Holes 61
Steel
    CPM 10V 52
    Cutting Large Sections 74
    O1, A2, D2, S7 52
    Stainless Steel 52
Stellite 52
Stresses 74
    Metal 71
Stripper
    With Knockout 143
    With Springs 142

T

Tapers
    See Dies-Taper
Templates 42
Test Specimens 47
Titanium 52
Tolerances 67
Tool and Die 109, 119
Tool Bits 46
Trail Off
    Wire 70
Types of Wire EDM
    Independent Four Axis 29
    Simultaneous Four Axis 28
    Two Axis 28

V

Various Wire EDM Applications 99

W

Wire
    Barreling Effect 70
    Corners 69
    Kerf 68
    Miniaturization 40

What Others Say

EDM Today, Jack Sebzda, Sr. Editor in Chief

"FINALLY...A comprehensive, professionally written book, all about Wire EDM is available to the EDM community!...

"The 'Wire EDM Handbook' puts a wealth of practical, to-the-point information at your fingertips. Written for people at every level of EDM experience, this professional, hard cover book, belongs in every EDM shop."

(EDM Today has ordered more than 700 copies.)

Society of Manufacturing Engineers

"You'll learn more about how wire EDM works, why it works, when it works best, and how it's revolutionizing tool and die making. Chapter topics cover profitable production tips, typical parts and part features, starting methods for edges and holes, accuracy and tolerances, and applications for die repair, punches, die sections and blanking."

EDM Technical Resource Directory

"A comprehensive source book on Wire EDM for designers, engineers, machinists, tool and die makers, mold makers and others in the metal machining field."

Barry R. Boman, Vice President of Kern EDM Supplies

"'Wire EDM Handbook' is must reading for those starting out in wire EDM....those designing or laying out wire EDM jobs, engineers, shop owners, or those contemplating a wire machine purchase."

Fabricators and Manufactures Association

"If you job EDM work out, are just starting with EDM, or are just curious about what it can accomplish, this is the book to read! Written for people at every level of EDM (Electrical Discharge Machining) experience, this professional, hard cover book belongs in every EDM shop."

Wire EDM Handbook is now being used as a textbook in U.S. colleges and technical schools.